COLLECTING and PRESERVING PLANTS for SCIENCE and PLEASURE

Ruth B. (Alford) MacFarlane

Illustrated by Jean Lynn Alred

ARCO PUBLISHING, INC.
NEW YORK

This book is dedicated with gratitude to Dr. Richard A. Giles, who opened the door.

I sincerely thank all those who shared with me their knowledge, their time, and their encouragement, with a special word for my husband, Richard, who endured some gray winter holidays while I worked like a mole completing the manuscript.

Published by Arco Publishing, Inc.
215 Park Avenue South, New York, N.Y. 10003

Copyright © 1985 by Ruth B. Alford MacFarlane

Library of Congress Cataloging in Publication Data

MacFarlane, Ruth B. Alford
   Collecting and preserving plants for science and pleasure.

   Includes index.
   1. Plants—Collection and preservation.   I. Title.
QK61.M32   1985   579   84-14593
ISBN 0-668-06009-3 (Reference Text)
ISBN 0-668-06013-1 (Paper Edition)

Printed in the United States of America

10 9 8 7 6 5 4 3 2 1

Contents

Introduction

We live surrounded by plants. Fortunately so, for *all* the world's food and much of its energy is based on plants, modern or ancient. Let us notice them, enjoy them. One way to do so is to collect and preserve them.

This book attempts to guide you through the steps of collecting and preserving plants. It is written with two purposes in mind: first, to provide information as to how to collect scientifically; and second, how to do so for ornamental uses. The techniques overlap. The same process that produces a carefully pressed and dried scientific specimen may be used to prepare plants for decorative display.

One important scientific use for which plants are collected is for a herbarium, a collection or library of preserved plants, used in research or teaching. Herbaria are often associated with colleges and botanical gardens, but many are maintained by individuals. The value of a herbarium specimen lies in the fact that the actual plant is there, not just someone's description or drawing of it, which may contain errors. The plant is never wrong. There it lies, itself, misnamed perhaps, but always capable of being taken out and restudied. Properly prepared and protected, such specimens may last hundreds of years. History, too, is written on the labels of those plants, the history of plants and people, of explorations and of plant migrations and utilization.

Besides being collected for use in the herbarium, plants are preserved as class projects, for research on particular groups, as voucher specimens for a certain study, and to use in teaching. Other educational uses for collecting may be as projects for Four-H, Scouts, senior citizen groups, and garden clubs.

Ornamental applications may be as pictures, wall hangings, arrangements in containers, plastic-laminated place mats, coasters, stationery, and many other attractive uses.

A good scientific specimen should be representative of its kind, not the smallest or largest (unless it is necessary to show the size range of a given plant). It should be relatively undamaged so that leaf shapes and sizes will be visible, and it should be complete, with root, stem, leaves, and *flowers* or *fruit*. It should be painstakingly prepared, with leaves and flowers uncrumpled, and with key characters displayed. Finally, it should be accompanied by accurate data describing exactly when and where it was collected.

Most of these stipulations would apply also when plants are used for decorative purposes, although in such cases usually the roots would not be needed. Selection would be made for effect, not for scientific value, and such accurate and complete data would not be needed.

There are some *restrictions* and *cautions*.

Poisonous Plants

Be alert for poisonous plants and know what to avoid.

Poison Ivy may be a shrub or a high-climbing vine, with aerial rootlets prominent and characteristic. It has three leaflets, the center one of which is set up on a stalk. In early summer it bears an attractive panicle of creamy white flowers. The fruits are small, white, berrylike

Poison ivy in fruit   *Leaf variation in poison ivy*

Poison sumac in fruit

drupes. In fall the yellow and scarlet leaves invite collecting, but they must be avoided.

Poison sumac is a shrub often so small as to appear to be a ground-cover plant, but generally about 8 to 15 feet tall (2.5 to 5 m), with compound leaves and white drupes (again similar to berries in appearance), which remain on the shrub all winter. It grows in boggy places where sphagnum moss is often to be found also, and where sodden ground may induce the collector to grab for the nearest support—though not the poison sumac one hopes.

After contact with either of the above plants, thorough scrubbing of the affected areas of skin with naphtha, or any soap, as soon as possible after contact, may help to avert outbreaks on less sensitive skins. Otherwise, consult a doctor.

Jimson weed is poisonous to some people. It is an upright plant, 3 to 5 feet high (1 to 1.6 m), with large lavender-white trumpet-shaped flowers, and ovoid spiny fruits. When the fruits are dead and dry, there seems to be no trouble.

Protected Plants

Because of the pressures of civilization, destruction of habitats, and overcollecting, especially for commercial purposes, some of our most beautiful plants are becoming rare and in danger of extinction. As a plant

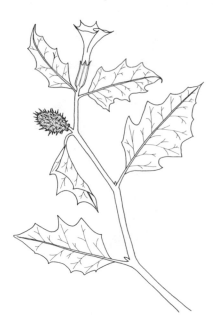

Jimson weed

lover and collector, do not contribute to this destruction. Become familiar with which plants are on the threatened and endangered species lists and avoid them. In most states, the departments of natural resources can supply lists of such plants and information about recent legislation concerned with plants. For the national list, write to:

U.S. Fish and Wildlife Service
Department of the Interior
Washington, D.C., 20240

Other organizations helpful in this regard are botanical societies, botanical gardens, and conservation groups.

Avoiding the protected plants, turn your attention to the common ones, which present an often overlooked beauty and diversity. Dandelions are lovely when pressed, and you are not likely to wipe out that species!

Lady slipper

Permission

Respect the property of others. It is important not to dig and gather plants without first getting permission from landowners. To some city dwellers, the wider stretches of land in the country seem to belong to no one, but the opposite is true. It is safe to assume that there is hardly a square foot of land in the world that is not owned by someone. As an example, I once planted daffodils on the roadside bank in front of my country home. Imagine my wrath when a passing motorist stopped to cut the "wildflowers."

Take the trouble to find out the name of a landowner and to ask permission, and explain the reasons for wanting to collect. Usually permission will be given. Often the owner will go further and lend you active assistance, perhaps even pointing out choice collecting spots. The same methods hold true in urban areas. You can find out the names of owners by simply asking at the nearest house, or by consulting township, county, or city records, or by looking in plat books available at government offices or in local libraries.

State and national parks may be closed to all collecting, in the attempt to save the wild plants for all the people to enjoy, but you might be granted permission to do limited collecting in certain parks by asking at the ranger station, or by writing to state or national headquarters.

In any case, do not contribute to the destruction of species. If you are looking at a huge patch of a plant, digging one will do no harm. If there are only a few, leave them there. Never take the last one.

And now, on to the methods.

1

Collecting

One reason for collecting plants, I suspect, is that it takes us out in the woods and fields, where we may breathe in the scent of leaf mold and grasses, feel the sun (and sometimes the rain) on our faces, and just enjoy being outdoors. Others may play golf; we collect plants.

The way we collect a plant will determine how useful or attractive it will be either as a scientific specimen or as a decoration, so from the very beginning it is important at each step to follow correct procedures for collecting and preserving.

Before setting out, assemble the materials for recording, digging, cutting, tagging, and carrying. The basic set, listed here, may vary depending upon the purpose of the trip. On an all-day field trip, or even a prolonged trip of several days or weeks, it is important not to forget anything, though for a trip to the backyard, simpler preparations will do.

Materials

1. For recording, use a small spiral notebook (3 × 5 inches) and a pencil (*not* a pen, in case the notebook gets wet), both hung on strings around the neck, to leave the hands free, or carry the notebook in a pocket.

2. For digging, a trowel or dandelion cutter is useful. Some people prefer a larger tool, like a mattock or spade. It is up to the individual to decide if it is worth the extra weight.

3. A jackknife or a pair of hand pruners will be useful for cutting twigs from wooden plants, or scraping lichens from trees and rocks.

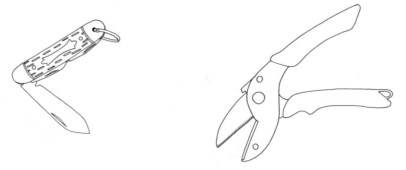

4. A sturdy tag will be needed for temporary numbering of the plants collected. String-tags are good, or scraps of paper with a high rag content.

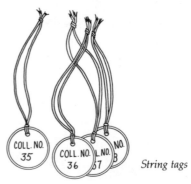

String tags

5. For carrying, take along plastic bags of various sizes. The 11 × 14 inch size (28 × 36 cm) is the most useful, but small sandwich bags (4 × 5 inches; 10 × 13 cm) are just right for mosses or other small plants.

Several larger bags will hold bigger plants or will transport the filled ones. Many people still use a vasculum, a cylindrical metal box hung by a strap from the shoulder. Lined with wet paper, it makes an excellent plant carrier.

Vasculum

6. A hand lens, also hung around the neck, will permit close examination of the plants, to be sure they are being collected in proper condition. Some flowers are very small and hard to see.

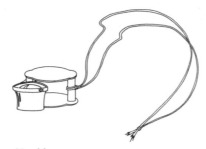

Hand lens

7. For organizing everything, and for carrying little extras such as vials or lunch, a fisherman's vest or canvas creel, or a knapsack, is useful.

8. For comfort, insect repellent will almost certainly be needed. In extreme conditions, a head net will make life bearable.

9. A camera is optional, but sometimes it will be necessary to photograph habitats.

10. For plants that must be pressed instantly, such as quickly wilting ferns, a lightweight field press will be useful. Usually a plant press will be left in the car or at the camp.

Certain collecting problems may require special equipment. Waterproof clothing may be necessary. Wading, or even diving, gear will be essential for reaching the sites of algae and other water plants. A rock hammer will chip off bits of rock on which lichens and mosses grow.

Techniques of Collecting

Before digging the plant, *record the data*. Include location, date, collection number, habitat, and miscellaneous observations.

Locality notes should guide a complete stranger to the spot 50 years later. Use landmarks, give mileages, names of roads, and nearby towns as exactly as possible, and always include the state, province, or country (and the county and township). If it can't be filled in on the spot, record enough so that the rest can be looked up later.

The *date* should include the day, month, and year.

Collection numbers should begin with "1" for the first plant collected and continue consecutively, a different number for each specimen, except for several taken at the same time from the same species. Some people like to start a fresh series each year, numbering them "84-1," "84-2," etc. Experienced collectors recommend that the collector just keep on counting, throughout a lifetime.

Habitat notes describe whether the plant grows in sun or shade; woods, field, marsh, water, etc.; soil type—sand, clay, gravel, humus, rock cleft, or other; wet or dry; associated plants; altitude if in the mountains; water depth if in water; and various refinements of the

"At your feet is a flowering plant"

above, that is, "open shade of oak scrub." Often there is a need to bring back specimens for people working with other plant groups. If complete and accurate notes are kept such specimens will be valuable.

Miscellaneous observations might give flower color, which often changes in drying, or any other notes that might be useful, such as flower, twig, or leaf odor, or a description of the plant if it is too big to take all of it, or, for example, "visited by many yellow butterflies." Mention anything that will not be obvious when the plant is dry.

Try to get plants undamaged by insects, disease, or accident (unless, of course, when making a study of insect damage or plant disease).

Dig the plant carefully, getting the whole root, which sometimes is very deep. Gently shake off most of the dirt, remove moss or dead leaves, wash off the roots if water is available, and place the plant in a bag along with a tag bearing the collection number. It is best to put only one plant in a bag, but if plants are doubled up, tag the plant itself to avoid confusion later. Use an indelible, fine-point felt pen on a twist-tie which can then be attached to the plant. Close the bag with a twist-tie or a rubber band (or simply tie it) after removing excess air to retard drying. If available, a few drops of water or a damp paper towel in the bag or vasculum will help to preserve freshness. If the plants are delicate, inflate the bag to protect the specimen from being crushed. In a vasculum, keep the roots at the same end so as not to get dirt on the flowers.

While collecting a plant, get a few extra plants or flowers to use in identifying. Fresh material is easier to use. Just be sure it is the *same plant*! Sometimes similar species grow intermingled.

On first field trips, collect only a few plants—five plants are enough in the beginning. They will all have to be pressed at the end of the trip. It is better to make more frequent trips and collect only enough to process carefully.

The foregoing describes the collection of a straightforward plant. There can be special problems.

Special Problems

Trees and Other Woody Plants

Trees are not only woody, they are *tall*. And they do have flowers, often of great delicacy and beauty, such as those of the American elm, one of the first trees to blossom in spring.

Obviously, the whole plant cannot be collected, so it must be described instead: height; diameter-at-breast height (DBH); manner of growth (erect, arching, vase-shaped, prostrate, conical, climbing, etc.); and the type of bark

Cutting a branch from a tall elm tree

(peeling, warty, platy, shaggy, fine-grained, smooth); as well as the color of the bark and anything else that may be distinctive. To estimate the height, I think my way up the tree, 6 feet (about 2 m) at a time.

If no branch is within reach to snip off a twig (not more than 12 inches or 30 cm long) with pruners or knife, long-handled pruners are available from farm-supply or hardware stores. Some of these collapse to a length of about 6 feet (about 2 m) and will even fit (barely) inside a compact car (an agile person can climb the tree). Be careful not to collect only juvenile or shade leaves, since they often differ from adult leaves as to shape, size, or texture. If possible collect a piece of bark.

Many woody plants (e.g., willows) as well as herbaceous ones, must be revisited for later collections of flowers, fruits, or leaves, in order to be accurately identified. Such specimens must be taken from the *same* tree or

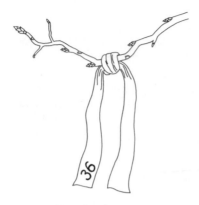

Marking flag on a bare twig

shrub. The only way to be sure, later, when exuberant summer leaves obscure the simple outlines of spring, is to mark the plant with a flag of some sort. A tiny tag will disappear among the leaves. A strip of old cotton sheet can be used as a marker, with the collection number written on it firmly in pencil. Tied to a branch, such a flag will remain legible for a full year and act as a guide back to the same tree or shrub.

The first specimen taken from that marked plant is assigned the collection number. The second is given that number plus "a" or "b" to show that it is from exactly the same plant. Sometimes both specimens are mounted together on the same sheet, and the different times of collecting explained on the label.

Marking flag on a leafy twig

Fruits

Fruits and seeds are very individualistic, very valuable to identification, and ornamental in arrangements.

Take a small piece of the parent plant as well, enough to show the leaf arrangement (opposite or alternate) and including the tip of the twig, along with the fruits or seeds. When gathering the fruits from the ground, be absolutely sure they fell from the same tree or shrub. Field notes should describe where they were found. Fruits and seeds are often carried by wind or animals.

Mosses, Liverworts, and Lichens

Lichens, although not closely related to the mosses and liverworts, are treated with them because of their similar methods of collection and preservation. The three groups usually consist of small and inconspicuous plants. Many, but by no means all, prefer damp places.

A variety of species may be found on a single large boulder, depending on the light and moisture available. At the base of the rock will be moisture-loving plants; at the top, those that resist drought. Acid or alkaline soils support different species, as do trees and shrubs on their bark, twigs, and even leaves. Some mosses and liverworts grow in water. Be sure to record the details of habitat exactly.

Also collect a sample of the substrate. Rock can be scraped off with a knife or chipped with a rock hammer. Take a bit of bark or twig, or some soil, although most of it will be gently washed off. Notes should record the color of the plant, the nature of the substrate, the distance above the ground, and the amount of light. If the kind of tree is unknown, bring back enough material to identify it.

In a dry climate, transport specimens in paper sacks or folded squares of newspaper, but in general plastic bags are better. Small cloth bags about 4 × 8 inches (10 × 20 cm) may be used. Place each species in a separate sack and enclose a stout, penciled tag bearing the collection

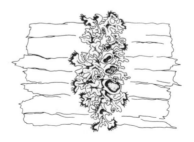

Lichen on bark

number. If there is any chance of the field notebook being separated from the specimens, it is a good idea to place the date, name of collector, and locality information on that tag also.

Take mosses only when they are bearing the spore capsules, which are generally raised noticeably above the rest of the plant. Liverworts are collected at any stage. On lichens, look for the cuplike fruiting structures—use of a hand lens may be necessary to find them.

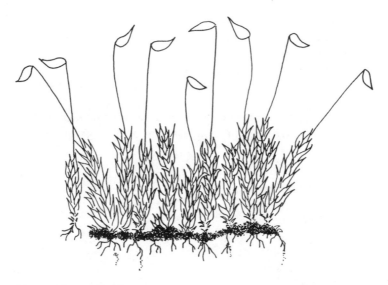

Mosses with spore capsules

Algae and Other Water Plants

Plants that grow in water do not have the stiffer supporting structures that land plants have, so when they are removed from the water they tend to collapse into a formless, often slimy mass. Transport them in containers along with some water.

Special data to record are *color;* water depth; flowing or still water; fresh or saltwater; whether the plant is immersed (under the surface) or emergent (protruding above the water); free-floating, epiphytic (growing on another plant) or attached to the bottom or other surface such as a bank; the nature of the substrate (muck, sand, gravel, rocks, wood); pH (acidity) of the water; tidal conditions, if marine; and odor.

Sample any algae over an area of several yards or meters, trying to get the various reproductive stages necessary for identification. Along the seashore, washed-up specimens are all right to take as long as they are fresh and not bleached or decayed.

Unless collecting on a lakeshore or ocean beach as part of a class or expedition that already has a permit, be sure to obtain permission. Some beaches are closed to all collecting; others require special permits.

Some algae grow in damp places on bark, leaves, or soil. Collect some of the substrate too and transport the algae in bags like any other land plant. Intertidal-zone collectors of marine algae usually place the specimens in plastic bags along with some seawater, then put the bags in a bucket, for stability. The bucket floats and keeps the load from being too heavy. Some bags have a special twist-tie closure.

Freshwater algae are usually microscopic in size (although in large numbers they can scum a whole lake), and they are collected into vials with a penciled label bearing the collection number slipped in. Some collectors preslice a 3 × 5 inch index card into narrow strips, almost all the way through, and tear off a strip as needed. Leave a little air space at the top of the vial.

When recording the data in a field notebook, take note of any peculiarities of growth. Is the alga sky-blue or blue-green? Does it look

 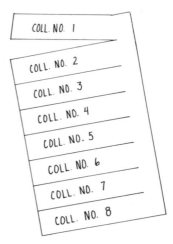

"cottony"? Does it look slimy, oily, stringy, or like green paint? Does it have a scent? (One alga smells like garlic.)

Fungi

Fungi contain no chlorophyll, the green coloring matter that enables most plants to manufacture their own food from chemicals in the air, water, and soil. Fungi grow as parasites, drawing their nourishment from living plants or animals, or as saprophytes on such dead matter as

Field mushroom

leaf litter or compost. Although most are colorless, some can be rusty, bluish, pink, black, cream-colored, or even bright red. They send out microscopic filaments, called mycelia, that grow upon or within the host, dead or alive. Sometimes these filaments intertwine to form a larger structure, the fruiting body, which is the familiar mushroom or other type of fungus that you see. Others appear as reddish or blackish spots on other plants. Rusts, for example, may form rust-colored patches on leaves or stems. Smuts form blackish bulges and lumps. You should collect enough of the host plant so it can be identified. If fungi grow on twigs, branches, or bark, cut off some of the woody material, too.

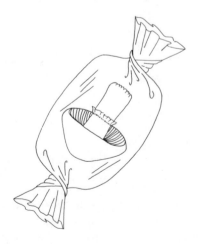

Mushroom wrapped for carrying

Fruiting bodies of the larger fungi may be cup shaped, spherical (puffballs), club shaped, repeatedly branched and coral-like, umbrella shaped (mushrooms and toadstools), or shelflike.They may be woody, corky, or fleshy. Wrap specimens of this type separately in pieces of waxed paper with the ends twisted shut, or even in squares of newspaper, and carry them in a basket to avoid crushing. Place a tag in with each, bearing your collection number. Use sturdy paper, as the tag will have to go through the drying process with the specimen.

Field notes are important. In addition to the collection number, the date, habitat, and locality notes that go on every specimen of whatever type, be sure to record (1) the color, size, and shape, since fungi may shrink or twist in drying, and the color may fade or change; (2) the odor; and (3) particular details, including the name of the host plant. Exact details of where it grows will help in identification later, such things as "on freshly cut white oak log," "on rotting box," "on standing dead tree," or "on living lilac shrub." Note the relative abundance: rare, common, or abundant.

Plants for Ornamental Uses

When plants are collected for decorative purposes, it is easy to think only in terms of the more obvious ones like strawflowers, teasel, or milkweed pods. But after a while, every flower, leaf, twig, fruit, or weedstalk will be assessed for its possibilities.

Each season of the year brings its opportunities to collect. Winter reveals the shapes of bare twigs, the texture of bark, and the seed heads of weeds thrusting up through the snow. The catkins of birch and alder wait, perfectly formed but contracted, until warmer weather brings release. Spring provides delicate flowers, especially those in wild woodlands, and the tiny new leaves of ash, oak, and maple. Summer produces abundance: sturdy zinnias, lacy ferns, feathery grasses, or the seed-laden stalks of dock changing from green to pink to coppery brown. Fall showers down bright leaves, nuts, and other ripened fruits.

Look everywhere for specimens, from the flower bed to the vegetable garden, from house plants to roadside weeds, from woodland and bog to lake and seashore. There are pieces of driftwood, contorted twigs, pine cones, twisted vines, flowers in a rainbow of colors, and fruit pits to consider. There is beauty in common plants, and it is possible to cultivate plants especially for decorative purposes.

In Appendix 1, there is a list of plants to grow or collect for drying, giving the stage at which each should be collected, their colors, parts to be used, method of preservation, and special notes.

Begin close to home, and begin simply: Flowers gathered in the yard can be processed immediately, thereby keeping their best shape and color.

Chinese lantern

Freshness is essential; wilted or faded flowers will not give good results.

The time of day and the plant's stage of maturity are most important. Go out in the early afternoon on a sunny day, after the dew has dried but before the flowers have begun to fade, never right after a rain. Gather only as many flowers as can be cared for at once. Most flowers should be taken when they have just reached their prime, but some should be gathered in bud (freesia, for example), or when just beginning to open (strawflowers). It is preferable to wait until the petals of mint-family flowers have fallen, then gather the stiffer sepals. Cut grasses when the panicles first spread, because later they may contract to a club shape or drop their seeds. Leaves taken early in the year will keep better color and be less likely to show insect damage. Select only perfect flowers and leaves, since every imperfection will be exposed in an ornamental arrangement. Sometimes it will be preferable to take the whole plant, but more often individual flowers, leaves, or fruits will be selected.

Follow the same collection techniques as for scientific purposes.

It isn't necessary to keep records as complete as for scientific collecting, but do keep a little notebook, telling when and where a plant was gathered, its original color, and other notes of interest. An excerpt from my own notebook reads: "No. 13, June 14, 1967. Downy Brome grass. Edge of garden, Beck Road. Dried upright in jar in kitchen. July 12. Sprayed [with clear plastic spray]. Stored in box. Silvery pale green. Good. Also hung some. Excellent. Pale silver green. Better than above. July 12. Sprayed. Stored in box."

It will take time to develop judgment and skill. Work slowly and carefully at first. Aim for perfection but be patient; skill and speed will gradually improve. After a while collection steps will be performed quickly, tucked in among other tasks. While picking garden beans cut a stalk of dock flowers or a bristly green head of amaranth, strip away the leaves, and hang it to dry immediately.

Herbs

Herbs make a special category of plants to dry. Some of them work well in ornamental arrangements. If it is convenient to grow herbs from seeds, follow the instructions printed on the seed packet or received with prestarted plants. Gather the part to be used—flower, leaf, or seed—on a dry sunny day, and process them as directed in Chapter 2.

Plants for Fragrance

Gather scented flowers or leaves about noon on a dry, sunny day, when the plant is in its prime. Directions for drying can be found in Chapter 2.

2

Preserving

The best time to press or otherwise preserve the plants collected is *immediately*, while they are still fresh. If a wait is unavoidable, put the tightly closed plastic bags in the refrigerator (not the freezer); most will hold for 24 hours. Leaving them any longer than that gives time for mold to grow or for petals to drop or other unpleasant things to happen.

Pressing

Pressing is the usual method to use for the higher plants; that is, those having flowers and seeds.

Materials

A plant press, blotters or newspapers, corrugated cardboard ventilators or driers, and some little weights or pieces of soft plastic foam for special problems will be needed.

A *plant press* is essentially a pair of wooden grids which may be strapped tightly about the pile of pressed plants. Each grid is 12 × 18 inches (30 × 46 cm), about the size of a half-sheet of newspaper. It may be purchased or constructed. If you are making it or having it made, a good material is oak, hickory, or ash strips, ¾ × 1 inch in cross section (2 × 2.5 cm), placed together in a lattice pattern, and glued and screwed or riveted at each intersection. Other materials may be used, for instance, pieces of plywood or perforated Masonite. Just be sure the press is sturdily constructed, yet not excessively heavy.

Binding straps for the press may be of canvas webbing. They can be 1 × 72 inches (2.5 × 183 cm) or perhaps shorter, ¾ × 48 inches (1.9 × 122 cm), with a thumb lock. A lot of pressure is going to be put on that press. The length gives room for expansion if many plants are collected, as on a trip. Pressing in a book is not a good practice. The plants dry too slowly, tend to lose their color, and seldom dry perfectly flat.

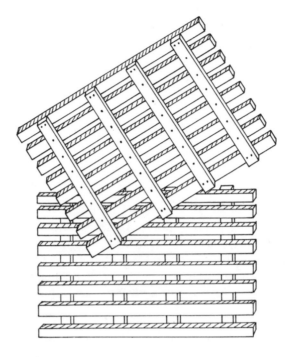

A plant press

Blotters 12 × 18 inches (30 × 46 cm) or folded newspaper of similar size will help to absorb moisture and hasten drying.

Corrugated cardboard ventilators or *corrugates,* also 12 × 18 inches, (30 × 46 cm) will separate the blotters and plants for air circulation. If corrugates are cut from boxes, have the corrugations running crosswise for best air flow.

Single sheets of newspaper or newsprint, about 11 × 16 inches (28 × 41 cm), when folded, are used to contain the plants. The reason for these dimensions is that this is nearly the size of standard herbarium mounting paper (11½ × 16½ inches; 29 × 42 cm). Plants that project from the pressing paper will not fit the mounting paper later.

Assemble materials so that they will be within easy reach, sit in a comfortable chair, and proceed with the pressing. Of course, under other conditions, a collector may have to work at the kitchen counter or before the open door of a van.

Procedure

Fold a *single* sheet of newspaper in half. On either the *outside* long edge or short edge, write the collection number and the date of collection. It doesn't matter where you write this information, but be consistent. If for some reason the pressed plants and the field notebook may become separated for too long, write all information on the newspaper margin.

Remove the plant from the plastic bag, leaving the rest of the plants covered so they won't wilt. Wash any remaining dirt off the roots and spread the plant out in a little less than the space of a half newspaper sheet. It is important to take plenty of time. This is the point where the appearance of finished specimens is determined.

Now, spread each leaf out flat. Turn over at least two leaves to show the underside, as the type of hairiness of that underside is often a crucial part of identification. Carefully spread open the flowers, making sure every petal lies smoothly. Try to press the flower in such a way as to best show important characteristics. For example, Queen Anne's lace has a single dark floret in the center of its white ones that should be displayed. Try to show the interior of at least one flower even if the important characteristics are unknown to you. Some of the flowers should lie with lower side uppermost, as did the leaves. *Every* leaf and petal must be neatly spread and unwrinkled; set high standards. I would be the first to admit that perfection is not always possible, but aim for it.

If the plant is too long for the sheet, fold it sharply in a V; if that is not enough, fold it into an N (see illustration, page 18); and if that is not enough, fold the plant into a W. If it is still too big, cut it and use only part, or press it in several sections, being sure to describe the original size of the plant in the field notebook.

Queen Anne's lace

Mounting a large plant, N-fold

This all sounds fine, but a plant to be pressed can develop a mind of its own. I keep on hand a set of small weights. They may be rubber stoppers, metal washers, metal nuts, little jars, salt and pepper shakers, flat rocks, or whatever lies at hand. As you get a leaf or flower spread, set a weight on it. When the whole plant is spread, remove the weights one by one, beginning at one edge, and simultaneously, with the other hand, bring the top half of the folded paper across to hold the parts in place.

Small strips of wet newspaper ½ X 3 inches (1.3 X 8 cm) are also effective in holding down thinner flowers and leaves. The wet strip will stick to the enclosing newspaper and does not have to be removed before pressing as do the weights. As the strips dry they fall away.

There! The plant is spread and captured between the folded sheet of newspaper.

Open the plant press, lay a corrugated cardboard ventilator on one grid, then a blotter or heavy thickness of newspaper on that, then the pressed plant in its paper. If heavy newspaper is used as a blotter, slide in the folded single sheet of newspaper, with plant inside, so that its *open* edge is against the *closed* edge of the thick folded newspaper (see diagram, page 19). This helps hold everything in and makes changing the heavier paper easier. If you are using blotters, lay a second one on top of the thin newspaper, then a second corrugated sheet, and you are ready to repeat the procedure with the next plant.

Sometimes plants wilted so much that pressing is difficult can be revived by putting them in their plastic bag into a refrigerator for a few hours. Some ferns respond to this treatment. Ferns are especially

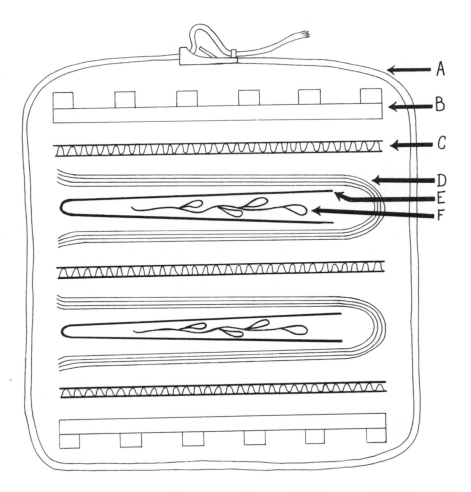

Loading a plant press—expanded view: A—Binding strap, B—Plant press grid, C—Corrugated cardboard ventilator, D—Thick newspaper used as a blotter, E—Single sheet of newspaper, folded, F—Plant being pressed

prone to curling and wilting in transit. It is still best to press plants before they wilt.

When all the plants collected have been pressed, place them, as described above, in a series of sandwiches in the plant press and strap the press shut, as tightly as possible. Having one person stand on the press while the other pulls the straps tight helps produce flat specimens. Loose strapping gives wrinkled leaves and flowers.

Set the press where a current of warm, not hot, air can flow up through the ventilators. A university herbarium may have a large drying cabinet with a heating element and a fan in the base, or a dryer built like a box with light bulbs or heating rods in the bottom. Use the facilities at hand—an open window, a place in the sun, on top of a register or flat-topped water heater, in front of a fan, or in the blast of air from something like a generator. An oven is too hot and causes browning.

Some collectors use a lightweight field press, consisting of corrugates and blotters between two pieces of Masonite or even just cardboards, strapped or roped together. They press their plants as soon as possible, right on the spot or perhaps at lunchtime or in the evening. In the field often several specimens must be placed between each pair of corrugates. Within 6 or 8 hours the collector will open the field press and rearrange the plants which by that time have become "tamed." This is the time to smooth out a petal or unfold a leaf, to be sure stems lie on top of leaves, not beneath, and that flowers are not covered. Some specimens can be improved a great deal. Others, aquatics for example, are too fragile to touch again. It is really safer to press the plants correctly in the first place if that is possible. As soon as possible, specimens are transferred to a regular press with fresh blotters and placed over mild heat.

A special press that I like to use, especially when pressing in quantity, is one tightened by means of wing nuts. These presses hold many plants under even pressure. Although a bit heavy, the press may be tilted on edge and placed over the heat source.

Most plants dry in 2 or 3 days to a week. After 24 hours, go through and *change* or *remove the blotters*. This is very important. The faster the plants dry, without high heat, the better the flower color will be retained.

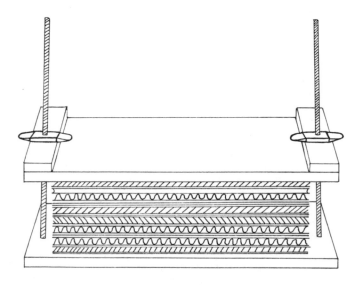

Screw-type plant press

After removing the blotters, return the press, strapped shut again, to the warm drying area.

If a plant feels cool to the touch it is probably not dry. When the plants are dry, remove them, still inside their single sheets of newspaper, which will remain around them throughout storage and until the plants are mounted, and store them flat in a protected place, usually in folders, about 15 plants in each folder, until they are to be mounted. Storing them in order by collection number will make working with them easier. Label each folder with a strip of paper protruding from the end, telling what lies within, to save much wasted time searching.

Special Problems

Now let's consider special problems in pressing.

Plants are often of uneven thickness with, say, thick stems and small flowers, or lumpy flowers and thin leaves. To do a better pressing job, pad up the thin areas with thick pieces of newspaper for the first 24 hours. This equalizes the pressure exerted by the press. I sometimes use such a thick pad to be sure a flower stays pressed open. The flowers of composites, such as the daisy, present a problem, with the thicker disk center and thin rays. One worker uses pieces of soft plastic foam over the heads, with excellent results. I have cut a "doughnut" out of several thicknesses of newspaper and set it over the flower head, letting the thicker center of the flower project through the hole in the doughnut. Your own ingenuity will show you ways to solve problems.

Trees and Other Woody Plants

The method of pressing twigs of trees and shrubs is closely similar to that of most herbaceous plants. The thickness of the twigs will give some trouble but may be compensated for, as described above, by padding with thicknesses of paper or foam. Here, again, be sure to turn over some leaves and press the flowers open to show their parts to best advantage.

Some twigs have just too many leaves to press well. I snip off some of the leaves, leaving the stubs of the petioles to show that they had been there. You need to keep branches from overlapping, yet still show the branching pattern. It is good practice to slice the twig diagonally to show the pith, because in some woody plants this is an identifying character.

Tropical trees may have *huge* leaves, nearly impossible to press. If for some special reason such a leaf is to be pressed, lay out the leaf on the

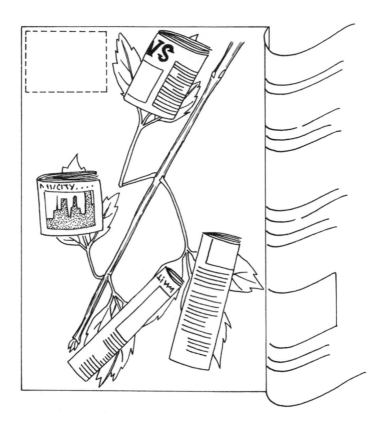

Woody twig with newspaper pads

ground or table and cut it into sections of a size to fit a herbarium sheet. Number or letter the pieces, to guide in reassembly, press, and mount.

Fruits

Some fruits are no problem at all. Small fruits, like berries, may be pressed still attached to their twigs. If they are very sticky or juicy press them between sheets of waxed paper, although they will not dry as quickly and they must be checked daily.

A good general rule for size might be that if the fruit has a diameter of 1 inch or less (2.5 cm), it can be pressed. Larger ones should be air dried or placed in fluid (see later in this chapter in the section on "Fluid Preservation"). Hickory nuts, for example, are dried in air for storage in boxes. Apples should be placed in a fluid preservative. Sometimes I have halved or sliced fruits, then pressed the slices, making a note on the sheet or label that this was done.

Mosses, Liverworts, and Lichens

Press these lightly by placing them in newspaper, then between blotters and corrugates, weighting them with books or other weights, not over 2 to 5 pounds (1 to 2 kg). Be careful with lichens as some may simply disintegrate if pressed. It helps to moisten them first, then press. Lichens and some of the mosses and liverworts may be dried in air for storage in boxes. If they are growing on other plants, press part of the host plant too.

Specimens to be used for dissection in classes may be stored in fluids.

Algae and Other Water Plants

Because these plants collapse when removed from the water, in order to press them you must first float them out in a shallow pan of water (saltwater if they are marine plants). Slide a piece of white watercolor paper or herbarium paper underneath, using a piece only large enough to take the whole plant. Arrange the delicate leaves and branches carefully with fingers, a narrow brush, or a blunt probe; then slowly withdraw the paper from the water, tilting it slightly to allow the water to flow off gently, and there you have it, a lovely tracery of color on the sheet.

Drying algae in a press is similar to drying land plants, *except* that the top of the plants must be covered with something to prevent their sticking to the enclosing paper or blotter. Either waxed paper or a piece of unbleached muslin is good. I like to lay my cards of algae within a single newspaper sheet, cover them with waxed paper or muslin, write the collection number and date on the outside margin of the newspaper, and insert that between the blotters and corrugates. I also write the collection number on each card bearing algae. Care must be taken at every step in the process to be sure that accurate records are kept.

Marine algae may grow to great size, in which case only a part of the plant can be pressed while the rest must be described. I have pressed one specimen that filled three herbarium sheets. Blue-green algae may be

Floating algae in a shallow tray

spread thinly over a sheet of paper and dried in a warm, shady spot, or covered with waxed paper and dried as above. Stoneworts such as *Chara* are pressed like seaweeds. They are very brittle and must be glued to the sheets later.

Algae are often preserved in fluids. Suitable solutions are given in the section on "Fluid Preservation," following. Note also the treatment recommended under "Glycerin Method."

It is possible to find oneself in a situation where supplies are unavailable for either pressing or fluid preservation, yet some specimens, especially seaweeds, have been found that may well be valuable to science. Here are two methods that will serve to preserve them during shipping to a center for study.

Lay the algae out to dry in a well-ventilated place. When they are almost dry, fold them up but do not pack them until they are completely dry, when they will be very brittle. Crate, enclose a label with each, and ship; or pack the well-drained algae layered with salt, like salt fish, crate and ship. However, only the coarsest types will survive this treatment.

Fungi

Fungi are rarely pressed. They are instead set whole in a warm place in trays or boxes or on wire screens to dry in air. In the field, camp stoves or lanterns or camp fires can supply mild heat. Do not cook them. Smaller fungi dry in about 3 to 4 hours at 100°F (38°C) in an oven; larger ones take several days. Keep their labels or tags with them at every step.

As the fungus dries, make a spore print. Cut a piece of waxed paper or smooth paper, preferably black, and place it beneath the fungus. During drying, spores of characteristic color and shape will be expelled. The spore print will aid in identification.

Sometimes fungi are dried in silica gel or some other dehydrating agent (see "Drying Agents"). Fleshy fungi are also put into liquid preservative (see the discussion under "Fluid Preservation"). If the fungus is on another plant, and is quite small, press it, host plant and all, just as you would any plant.

Plants for Ornamental Uses

If a flat arrangement of plants is planned, they must be pressed flat. The instructions given above for pressing scientific specimens will produce the best results here also, except that the root will probably not be needed. It may be necessary to separate the flowers and the leaves from the stems in order to give careful attention to each one. They can be glued back together later.

Modify the technique for pressing as follows. When it is desired to keep some of the plant's three-dimensional quality, or to make the leaves

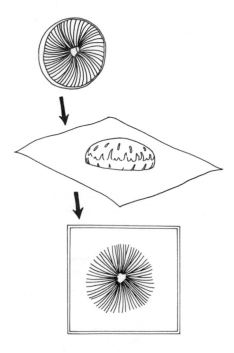

Making a spore print

a bit crinkly, lay four thicknesses of newspaper on the floor, lay fronds or sprays of leaves flat on this pad, cover with an additional four sheets of newspaper, add more leaves, more sheets of newspaper, and so on, then weight the whole pile and wait 1 month. Some foliage to press in this way is that of poppy, peony, vitex, cleome, and ginkgo, or autumn leaves.

Drying in Air

Many of the plants collected for ornamental purposes may simply be dried in air. Cut the stems to the length desired, neatly strip away leaves and any imperfect parts, tie the plants together tightly (stems will shrink) in small bundles, and hang upside down in a dry, airy place like an attic or closet. Tansy and dock and the fruits of mustard are among those to be dried this way, as well as most grasses. If the plants are bulky or look as though they may become entangled, hang them singly. Marigold, rose, peony, and hydrangea should be hung separately, one stalk to a string. Color retention is best if plants are dried in the dark. Some flowers hold their color better if they are placed with their stems in water a day or two before drying (pomegranate bud, yarrow, and ageratum). If dust, insects, or spiders and their webs are a problem, you can cover the plants with a paper bag. (A plastic bag would delay drying and permit the plants to mold.) I especially protect herbs this way. Some writers recommend drying herbs in a shady, not a dark place, so hanging them in a brown paper sack in the kitchen is just right.

For drying many plants, string clotheslines or set up a drying rack from which to hang the bundles. Just a few plants may be tied to a coat hanger and dried in a closet. Eight or ten days should be long enough to dry most plants. To allow for shrinkage, dry twice as much material as you think you need.

If you wish the stems to curve, stand the plants in a container or even fasten them to a curved wire while they dry. Hang vines and permit them to assume natural curves, or wind them about a support. Very delicate plants, such as leeks or polygonum, should be set upright in a fruit jar. Pinecones, nuts, twigs, and dryish fruits may be stored in boxes to dry.

Although leaves are usually preserved with glycerin, when they are to be used in vase arrangements, drying some in air will achieve unusual forms. Heavy, waxy leaves (orchid, rubber plant, magnolia, castor bean, aspidistra, or oak) lend themselves to the air-drying method. The leaves of canna, both green and brown, are excellent for drying. Try rolling some of them around a cardboard core or a rolled-up cloth. Vegetable leaves—green and purple cabbage, flowering kale, beet, and horseradish —will dry in dramatic shapes. Beech leaves gathered in spring will remain green as they dry; fall-gathered beech leaves will turn tan, and those taken after a hard freeze will turn white. Maple leaves should be cut early in color. Garden foliages such as artemisia, dusty miller, and woolly lamb's ear will retain their soft gray tones.

Watering

This is a variation of drying in air. The leaves should be large, strong in texture, and have long stems. Cut the stems at a 45-degree angle and place them in 2 inches (5 cm) of water in a tall container. In about a week, the water will have evaporated and the leaves will have begun to turn color. Leave them until they are completely dry. Calla, castor bean, dracaena, galax, lotus, and shallon will dry in this way.

Some plants, among them the heads of cattail, the seed pods of milkweed, or the catkins of willow, tend to go on drying, soon disintegrating and filling the air (and your house) with fluffy seeds. To avoid this, dip the plants in (or spray them with) a solution of cut shellac, 1 part of full-strength liquid shellac to 3 parts of shellac thinner (denatured alcohol), and hang them to dry.

NOTE: If the shellac has already been cut or diluted, adjust the proportions accordingly.

To produce a particularly delicate and beautiful addition to dried bouquets, try spraying the heads of goatsbeard (*Tragopogon*) with hair spray, thin shellac, or varnish; the plant resembles an oversized dandelion.

Fragrant Materials

To dry fragrant materials, which should have been cut in the early afternoon of a clear, sunny day, first place the stems in a small amount of water overnight. Next day, dismember the flowers. Spread them on absorbent paper in a warm spot, never in sunlight. Shuffle the petals or leaves daily for 4 or 5 days. Mint and rosemary leaves and the flowers of rose, carnation, geranium, heliotrope, honeysuckle, lavender, lilac, and spice pink are among the best to dry for fragrance.

Gourds

Gourds require special pains if they are to dry properly. Cut them from the vine, leaving a piece of stem. Wash *gently* to remove dust and garden dirt. Immerse each gourd in a household disinfectant solution such as Lysol or chlorine bleach, mixed in the proportions recommended on the container for ordinary use. Prick a tiny hole in each end of the

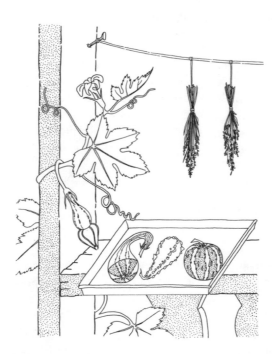

Gourds

gourd with a hatpin, for ventilation during drying. Attach a string to the stem and hang the gourd to dry for several months. Gourds are dry when the seeds rattle when you shake them. You may also lay gourds on a raised, perforated surface or on a screen in a cool spot. Turning them several times will hasten drying. When they are thoroughly dry, protect the surface with wax or with shellac cut 1 to 1 with shellac thinner.

Citrus Fruits

Citrus fruits (lemon, lime, orange, and tangerine) as well as pomegranate can be dried the same way as gourds, on a rack in a cool spot. Pick out firm, not overly ripe, blemish-free fruit. Wash and dry them (no disinfectant necessary) and lay them on a rack. Even if the fruits get a little soft, just wait. They will harden. They require no finish. The skin turns reddish brown with a rough texture. Pomegranate takes 3 months to dry and may shrink a little, but will retain its vivid color.

Drying Agents

To preserve flowers or some fungi or fruits in dry, three-dimensional condition, often wonderfully lifelike, try using a dehydrating agent. Plants prepared by this method are simply buried in the agent and left until dry. They often retain both natural color and shape. This is a very old method, but there have been some recent refinements.

There is a choice among several different agents: sand, borax, salt, silica gel or one of its commercial forms, pumice, cornmeal and other granular cereals, even cornstarch, powdered sugar, fuller's earth, or alum. A combination of two or more of these usually works best. Whichever drying agent is used, be sure it is clean, dry, and sifted free of any lumps or foreign materials.

Agents can be used over and over, by keeping them sifted clean, by drying them again by heating in a moderate oven, and by storing them in moisture-proof containers.

Sand

The oldest, and slowest method utilizes plain, clean, sifted, dry beach sand. You can purchase bags of sand at a lumber supply company.

Borax and Sand

One part of borax to 2 parts of sand makes a good ratio. Borax is easily obtainable from the laundry-materials department of the supermarket or grocery store.

Powdered Pumice and Yellow Cornmeal

Mix together equal parts of the two ingredients, and add 1 tablespoon (14 g) of *uniodized* salt per quart (946 ml) to keep the color of the plant brighter. This is a favorite combination.

Borax and Yellow Corn Meal

Mix 1 part of borax to 1 part of yellow cornmeal and add 1 tablespoon (14g) of *uniodized* salt per quart (946 ml) as above. Or use a combination of 5 pounds (2.25 kg) of bolted (finely sifted) yellow cornmeal with 25 ounces (709 g) of borax. Dry the mixture after each use for an hour at 150°F (71°C), and store the mixture in an airtight container.

Salt

Pure, *uniodized* salt dries plants rapidly and is especially good for flowers like delphinium and larkspur. Leave the flowers in the salt only 8 hours. Keep them moderately warm during the entire drying time, not above 100°F (36°C) or it will alter the color of the flowers.

However, in a book written in the nineteenth century telling of life in the fifteenth century, the author describes a method used at that earlier time. Roses gathered in summer were buried carefully in "bay salt" (salt obtained by evaporating seawater) in clay flowerpots, covered tightly, and set away in a cellar until Christmas. They were then taken out, set in warm water, brushed with a little wine, and used to make garlands to be worn on women's heads.

Silica gel

Silica gel is a modern drying agent sold by biological supply houses under that name or by florists under various trade names. Buy the finest grade, about the consistency of granulated sugar. Be careful not to inhale the dust when using silica gel. Remember, if it will withdraw moisture from a plant, it will not be good for the lungs.

SILICA GEL SHOULD BE USED ONLY BY ADULTS OR UNDER THE SUPERVISION OF AN ADULT. IT IS RECOMMENDED THAT A FACE MASK BE WORN WHEN USING THIS SUBSTANCE.

A borax-and-sand mixture, although slower acting, is much safer.

Indicator crystals of cobalt chloride ($CoCl_2$) are either included in the silica gel, or sold separately to be mixed with it. These are blue in color when the silica gel is dry and ready for use. They turn pink with moisture. Heating the mixture in a flat tray in a mild oven (250°F or

126°C) for about 30 minutes will dry out the silica gel and turn the crystals back to blue.

This is a relatively fast-drying agent, with a drying time of about 12 hours for fungi. Various foliages can be dried in silica gel, as can most flowers. Because of the rapid drying, foliage and flowers tend to keep their color. Autumn leaves, meadow rue, and fern are especially attractive dried in this way. Experiment with drying times, beginning with a check on your plants after only 24 hours.

Other Agents

Cornstarch, powdered sugar, fuller's earth, and alum are some other agents that may be tested. Of all the choices given, the drying agent that is least expensive and most readily available is probably the one to settle on.

Granular cereals like Wheatena, Cream of Wheat, or cornmeal, added to any other agent, will coarsen the mixture and help to keep it from sticking to the plant.

Method of Drying Plants in a Dehydrating Agent

Because borax and sand make an easily obtainable, middle-of-the-road drying agent, I shall use it as an example when describing the general method of drying flowers in an agent.

Gather flowers when they are just freshly opened, but not wet from dew or rain. Process them at the peak of readiness; inferior flowers will produce an inferior product. Strip away the leaves, keeping a stem about 6 to 8 inches (about 15 to 20 cm) long for single flowers or flower heads, or cut a 12- to 18-inch (about 30 to 45 cm) stalk for a plant like delphinium. Set the cut flowers in 1 to 2 inches (about 2.5 to 5 cm) of water in a glass. They will suck up water, which will prevent them from wilting during the first stages of drying.

Use a metal or plastic box, or a heavy cardboard box—a shoebox for a few small flowers or a heavy corrugated cardboard one for a larger number. The box should be large enough to accommodate the plants without crowding. Pour a layer of borax and sand into the box to a depth of several inches (up to 10 cm). Set the flowers on that layer, top down, then carefully sift more agent around and over each flower until all of them are completely covered except for the stems, and the drying agent has been sifted into every crevice. The flower heads should not overlap or touch. Some people do cover the stems as well, in which case allow for at least an inch (about 3 cm) of agent below the box cover.

Label the box with the name of the plants within and the date of burying. If they are scientific specimens, be sure that collection numbers are attached to each. Set the box away in a dark, dry place for about 1

Plants drying in dehydrating agent

week. Users disagree as to whether the box should be covered. Use your judgment. If the surrounding air is dry, leave the box uncovered; if moist, it is better to cover it. The air temperature should be moderate in most cases, neither too warm nor too cold, about 70°F (25°C).

At the end of the drying time, *gently* pour off the drying agent. If the plants are still flexible, they are not dry and should be left in for a longer period. Carefully take out the flowers one by one, tap them gently to remove any excess drying agent, and brush them lightly with a soft camel's hair brush, perhaps slightly dampened.

Although the instructions given in the preceding suggest placing the flowers top down on the drying agent, this must be modified for particular cases. A whole stalk of flowers should be laid horizontally, then covered very carefully with the agent. If flowers seem likely to be crushed, set them face up, sifting the agent in between the petals. Construct a drying box by punching holes in the bottom of a cardboard box, inserting flower stems through the holes, covering the flowers with agent, and setting the whole box up on blocks so as not to bend the stems. A tray set beneath will catch what little drying agent trickles through the holes.

Drying times will vary with the agent used and the kind of flower. If pure borax is used, for example, leave daisies or jonquils in the agent only a day and a half. If left longer, they may develop brown spots. If pure sand is used, leave most plants in for about 2 weeks.

Here again, it is a good idea to keep careful records: when a plant was collected, both date and time of day; date of burying and of removal; color before and after drying; agent used; and any other general observations.

Flowers dried in this way will be very brittle and must be handled with great care. If they do break, they can sometimes be mended by gluing them back together. To restore the velvety texture of flowers such as pansies and snapdragons, moisten a soft brush with vegetable oil and gently brush the petals.

The stems of daisies and zinnias and similar flowers tend to be too fragile after drying to hold up the weight of the heads. Get around this by inserting a hooked florist's wire into the center of the flower head, which has been removed from its stem, and drawing it down through until the hook is buried among the disk florets. Do this before drying the flower. A second method is to push two slender wires into the base of the flower at right angles to each other, bend the wires downward, and twist them together. Later, carefully wrap such wires with green florist's tape.

Some of the plants that dry well in dehydrating agents are colorful autumn leaves of dogwood, hickory and sweet gum; flowers of bleeding-heart, candytuft, Canterbury bell, daisy, delphinium, deutzia, pink dogwood, narcissus, pansy, peach and plum sprays, lilac, lily of the valley, marigold, pompon chrysanthemum, flowering quince, snapdragon, stock, sunflower, violet, all roses double and single, and zinnia. You may wish to dry keepsake corsages or bouquets. Our great-grandmothers dried their wedding bouquets in this way, and kept them under a glass bell jar in the parlor.

Some plants do not cure well, among them cymbidium orchid, petunia, nicotiana, gloxinia, and primrose. A natural stickiness makes it

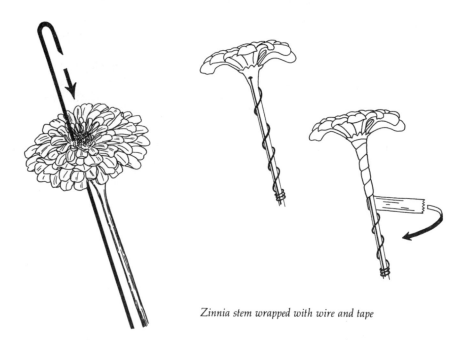

Zinnia stem wrapped with wire and tape

Zinnia with hooked wire
reinforcement

difficult to clean off all of the drying agent. Some plants are prone to reabsorb moisture quickly: these include most types of nasturtium, clematis, geranium, and evening primrose. This tendency may be counteracted by spraying them with a clear plastic spray after drying.

The possibilities are many, however. It is important to experiment and to keep precise records of experiments.

Fluid Preservation

For certain uses, or particular kinds of plants, storage in a liquid preservative is best. Flowers for dissection, fleshy fruits, algae, fungi, or any plant in which the three-dimensional shape is important may be stored this way. In most cases the preservative will remove the color.

Some of the more common fluids are listed in the following. Ethyl alcohol, formalin, glycerin, and acetic acid can be obtained from chemical supply or biological supply companies.

FORMALIN AND ACETIC ACID ARE SUBSTANCES TO BE USED WITH CAUTION AND ONLY BY ADULTS.

For the plant collections of youth groups it is best to use ordinary

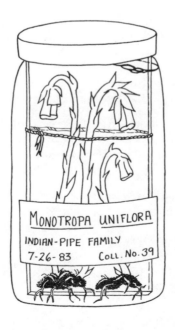

Indian pipe preserved in fluid

rubbing alcohol, a substance also useful for scientific collecting in emergency situations where the other materials are unavailable. Rubbing alcohol can be purchased in drugstores and supermarkets, but it *can* cause serious gastric disturbances if ingested.

Ethyl alcohol is the same alcohol found in alcoholic beverages, but usually a substance has been added to commercial and scientific supplies to make them unfit for drinking.

Please note that commercial formalin is sometimes referred to as 100 percent formalin, although it is actually about 38 percent. Treat the 38 percent formalin like 100 percent when mixing formulas.

1. *70 percent ethyl alcohol* (70 parts 95 percent ethyl alcohol to 30 parts distilled water). Good all-purpose preservative. Add 0.15 ounce (5 ml) of glycerin per 3 ounces (100 ml) of 70 percent alcohol to give added protection should the specimen dry out accidentally.

2. *50 percent ethyl alcohol* (50 parts 95 percent ethyl alcohol to 45 parts distilled water). Use for delicate freshwater algae. Add 0.15 ounce (5 ml) glycerin per 3 ounces (100 ml) of 50 percent alcohol.

3. *10 percent formalin* (10 parts 38 percent formalin to 90 parts distilled water, or sea water if a marine plant). Algae, or other plants, immersed in this, will keep indefinitely and may be used for preparation of microscope slides later.

4. *Six-three-one* (6 parts distilled water, or seawater if marine; 3 parts 95 percent ethyl alcohol; 1 part 38 percent formalin; plus 0.15 ounce (5 ml) of glycerin per 3 ounces (100 ml) of preservative). This is an excellent preservative for most plant materials, and will permit use of the tissues later for making microscope slides.

5. *FAA or formalin-aceto-alcohol* (50 parts 95 percent ethyl alcohol; 40 parts distilled water; 5 parts glacial acetic acid; 5 parts 38 percent formalin). A standard preservative for plants. useful in microtechniques also. There are variations of this formula for special uses.

For particular situations, you are advised to look further in books on biological techniques. Three good ones are:

Galigher, Albert E., and Eugene N. Kozloff. 1971. *Essentials of Practical Microtechnique,* 2d ed. Philadelphia: Lea & Febiger.
Johansen, Donald Alexander. 1940. *Plant Microtechnique.* New York: McGraw-Hill Book Co.
Knudsen, Jens W. 1966. *Biological Techniques.* New York: Harper and Row.

Sometimes in the field, where conditions are less than ideal, you must improvise. Measure formalin at the rate of 1 teaspoon of 38 percent

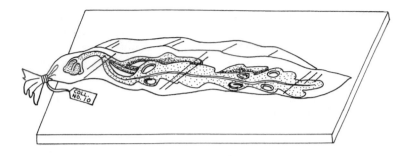

Glycerined algae

formalin to 1 ounce (30 ml) of water. If alcohol is used, use an amount equal to the specimens and the water they lie in. In desperation, use vodka or whiskey as an alcohol source, and hope that no one throws away your specimens and drinks the fluid. (That has happened.)

Glycerin Method

Impregnation of the plant tissues with glycerin can produce some special effects. Glycerin may be obtained from a drugstore or a biological or chemical supply house.

Algae

Large seaweeds, which ordinarily rot very quickly, may be washed gently in seawater, then soaked for a week in a solution of 1 part glycerin, 1 part 95 percent ethyl alcohol, and 2 parts seawater. Some people prefer a solution of 1 to 1 to 1 of those three substances. The seaweeds are removed from the solution, drained, and stored in closed plastic bags. Add a few crystals of phenol to each bag to prevent the development of mold. Phenol may be obtained from a biological-supply company.

CAUTION: PHENOL, OTHERWISE KNOWN AS CARBOLIC ACID, IS A POISON THAT IS ABSORBED THROUGH THE SKIN. YOU SHOULD WEAR RUBBER GLOVES WHEN HANDLING IT. USE PHENOL *ONLY* UNDER THE DIRECTION OF SOMEONE WITH A KNOWLEDGE OF AND RESPECT FOR ITS PROPERTIES. INEXPERIENCED STUDENTS, TEACHERS, OR AMATEUR NATURALISTS SHOULD *NOT* USE PHENOL.

The algae will retain their slimy–leathery texture for quite a long period, and make very interesting demonstration specimens, especially for students who may not have had an opportunity to observe the living plants. If they dry out, soak them in 1 part glycerin to 1 part water for a few days, *never* in just water alone, or they will fall apart.

Plants for Ornamental Uses

Infiltrating leaves with glycerin will produce flexible, long-lasting materials, often unusual in appearance. This can be done by standing leaf-bearing branches or stems in a solution of 2 parts of water to 1 part of glycerin, or by immersing the leaves alone in a 1-to-1 solution of glycerin and water.

Gather plants or small branches, wash them in cool water, remove the defective parts, and pound the lower 2 inches (about 5 cm) of the stems to crush them. This looks brutal, but it will split the bark and fibers and increase absorption.

Stand the stems in a jar containing the water–glycerin solution. Let them stand until absorption is complete, about 2 weeks for most plants. Russian olive takes about 3 to 5 weeks. When the plants are ready the glycerin begins to ooze from the leaf edges or the leaves have changed color. The method works best during the hot summer months. Occasionally, wipe the leaves with a damp cloth or with glycerin. During this process, the water in the plants will gradually be replaced by glycerin, so that they will not wilt during storage or use.

Glycerin gives a smooth, satiny look to most leaves. The color change is often to bronze or olive-green, which can be very attractive. Russian olive leaves, if left long enough, will turn bright yellow. Peony leaves acquire a silvery green or tan color. Barberry, gathered in spring, turns bright red when glycerined; gathered in fall, it turns brown. Some leaves retain their own texture and color—beech, crab apple, plum, forsythia, and almost all bronze-red foliage.

Whole leaves may be immersed in the 1-to-1 solution for comparable periods of time. Burdock, hollyhock, lily of the valley, and ivy are a few of these. Upon removal from the solution, the leaves are placed between sheets of newspaper and set beneath moderate weights. This blots up the excess glycerin. Leaves are then stored or used. Glycerined materials will last indefinitely.

The glycerin–water solutions can be used over and over.

This method often produces surprises. If a careful record is kept of when the plant was gathered, of just how and how long it was treated, and of the results obtained, it makes it easier to duplicate a desired result. Experiment and keep trying. You will gradually learn what works best for you.

Skeletonizing

Sometimes as you scuff through fallen leaves, you will come upon some that have been skeletonized. The covering and flesh of the leaf have been stripped away, leaving only a delicate lacework of veins.

To duplicate and speed up nature's slow process, try the following: Boil leaves for 1 hour in a solution of lye soap or in a solution of 1 teaspoon (about 5 cc) of baking soda (bicarbonate of soda) to 1 quart (946 ml) of water. Let the leaves cool in the solution. Spread them out on a newspaper and *carefully* scrape off the fleshy part on both sides with a dull table knife. Bleach them for an hour or so in a solution of 2 tablespoons (about 28 ml) of household chlorine bleach in 1 quart (946 ml) of water. Rinse them in clear water. Gently blot up excess water and press the leaves for 24 hours. They will appear white, misty, lacy—ghosts of their former selves. They will work well in flower pictures.

This is not an easy process. Many kinds of leaves simply disintegrate. The ones that work best for me are leaves with a waxy surface, like English ivy, or the leathery leaves of trees in fall. Experiment will show you the best ones for you.

When glycerined material gets old it naturally skeletonizes. You can hasten the process by stacking the leaves closely against one another, wrapping them in waxed paper which holds in their moisture, and storing them in a drawer or box for several months.

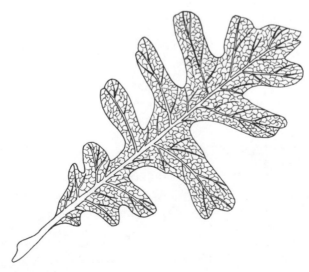

Skeletonized oak leaf

Freezing

Flowers and other plant structures for dissection in botany classes may be quick frozen for use during the winter months. Gather flowers representing particular families. Place them in plastic bags, label them plainly, seal, and store them in a freezer. When thawed they will be limp and not as good as fresh flowers, but will be better than dried or fluid-preserved flowers.

3

Identifying

There are many, many plant species out there, as you will learn when you begin to search. Your new awareness will help you to discover kinds of plants close to home that you never suspected were there. But you need a guide.

Books and manuals listed in this chapter range from those for amateur collectors to those for the taxonomy student, as well as some more specialized books. This list will be heavily weighted toward the northeastern United States, for that is the area with which I am most familiar. For manuals dealing with different parts of the country, inquire at park and natural area headquarters, botanical gardens, local bookshops or libraries, or college bookstores. Those interested in scientific manuals should inquire at university botany departments. Scientific supply companies also sell guidebooks. Federal government publications are yet another possibility.

Many of the books listed have glossaries. A specialized vocabulary is very useful to plant collectors. For example, it is better to use the single word *villous* in a plant description, rather than to say each time, "bearing moderately dense, long, soft, often curly hairs, more or less erect but not necessarily straight."

Most of the books listed will have some sort of *key*, the term applied to a series of choices that will lead, step by step, to the name of the plant at hand. Most keys are *dichotomous*, that is, each step consists of only two choices, but a few keys have three or more choices at each step. Read the introduction to the book to learn the general plan of the book and how to use the keys. Some field guides group plants according to flower color; others use picture keys.

Always read both or all of the choices given at any level of a key with great care before deciding which one to take, and consult the glossary if necessary. Write down the choices taken in abbreviated form. This makes it less likely that a step will be skipped.

The plant name will appear at the end of a statement and should provide identification of your specimen. Turn to the description under

that name in the manual and verify every characteristic. If these coincide
with the collector's own observations, and work has been done accurately,
the plant is identified.

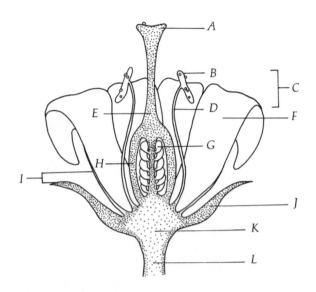

Diagram of a flower (adapted from D.C. Jackson). A—stigma, B—anther, C—stamen, D—
filament, E—style, F—petal, G—developing seed, H—ovary, I—perianth, J—sepal, K—
receptacle, L—peduncle.

Perhaps the best way to learn to use a key is to construct one. The
sample key is based on a collection of twigs gathered from six different
trees, all evergreens, and each twig bears a cone. Lay out the twigs and
observe their similarities and differences. Because most keys use the
metric system for measurement, only metric units are used in this
sample key.

Three of the specimens have relatively short needles, growing
closely along the twig. Three have longer needles, clustered into little
bundles. These two groups can represent the first grouping:

1-A. Needles scattered along the twig
1-B. Needles in bundles

Look at those twigs included under 1-A. How do they differ? One
twig has stiff, sharp-pointed needles. The other two have flat, soft
needles. If one of the stiff needles is broken, it is four-sided in cross
section. This provides the second division:

2-A. Needles stiff, sharp-pointed, four-sided in cross section
2-B. Needles soft, flattened

Of the two twigs with flattened needles, one has small, 6- to 15-mm-long needles, and little cones 19 mm long, that hang down on the underside of the twig. The other has needles 13 to 31 mm long, and cones 5 to 10 cm long that stand upright on the twig. This provides another choice:

3-A. Needles 6 to 15 mm long; cones 19 mm long, pendulous
3-B. Needles 13 to 31 mm long; cones 5 to 10 cm long, erect

That takes care of three twigs. The three remaining, placed under 1-B, have long needles in clusters. Look closely at those clusters. One twig has needles in groups of *five*, and cones 15 cm long. The other two twigs have needles grouped in pairs. The choices can be expressed as:

4-A. Needles in groups of five
4-B. Needles in groups of two

The five-needle specimen is set to one side as a 4-A. Look at the remaining pair of twigs. One has needles about 2.5 cm long and cones about 4 cm long, tightly closed and twisted. The second has needles 10 to 15 cm long, and cones that are 5 cm long, their scales spreading open. Your last pair of descriptions will be:

5-A. Needles 2.5 cm long; cones 4 cm long, twisted, tightly closed
5-B. Needles 10 to 15 cm long; cones 5 cm long, cone scales spreading

It is now possible to write the whole key to the six twigs. If they were gathered in the wild forest in Michigan's Upper Peninsula, for example, six of the most common conifers of this region will have been collected. Here is the way you would set down the key:

1-A. Needles scattered along the twig ....
    2-A. Needles stiff, sharp-pointed, four-sided in cross section ....
    .... *Picea* spp. Spruce
    2-B. Needles soft, flattened ....
        3-A. Needles 6 to 15 mm long; cones 19 mm long, pendulous
        .... *Tsuga canadensis* (L.) Carr. Hemlock
        3-B. Needles 13 to 31 mm long; cones 5 to 10 cm long, erect
        .... *Abies balsamea* (L.) Mill. Balsam Fir
1-B. Needles in bundles ....
    4-A. Needles in groups of five .... *Pinus strobus* L. White Pine
    4-B. Needles in groups of two ....
        5-A. Needles 2.5 cm long; cones 4 cm long, twisted ....
        .... *Pinus banksiana* Lamb. Jack Pine

5-B. Needles 10 to 15 cm long; cones 5 cm long, cone scales
spreading .... *Pinus resinosa* Ait. Red Pine

Notice that after choice 1-B was made, unless there had been a mistake and the key had to be started over, it would be unnecessary to go back to any part of the key above that point. Notice also that the key went only to *Picea* spp. in step 2-A. There are several common spruces; *Picea* spp. means the genus *Picea* (spruce), one of several possible species.

This key will help in the use of a tree identification manual, but a manual will contain many more species listed and more choices to be made.

To avoid destroying any important characteristics, I measure my way into the flower, first measuring and recording the dimensions and the shape of the sepals, then the petals, the stamens, and finally the ovary, which must often be cross-sectioned. After identification is completed, unless flowers are abundant, I carefully save the flower pieces between two small squares of waxed paper and place them in a fragment folder to be mounted with the plant.

If flowers are removed from a mounted herbarium specimen for study, a researcher must be even more careful to preserve the dissected flowers (see page 103).

Not all collectors will wish to carry their studies to this degree. For a beginner, learning to know the plants by their common names will be enough—to tell an ox-eye daisy from a black-eyed susan, for example. In the following list are flower identification books that will aid in identification often just by their outward appearance, without the use of the microscope. Although they are recommended for the beginner, these books are useful to the advanced student and amateur naturalist as well.

Ideally, a plant will be identified while fresh material is still available.

In practice, often the plant is pressed and dried, perhaps for years, before that can be attempted. Most of the characteristics of stems and leaves can be observed right on the whole pressed plant. Flowers, however, must be dissected. Formerly, dry flowers were boiled gently for 1 minute to soften them, but soaking them for a few minutes in water containing a drop of detergent will soften the tissues, as will boiling in water or soaking overnight in sodium hydroxide.

Plant Identification Books for the Beginner

Craighead, John J., Frank C. Davis, and Ray J. Davis. 1974. *A Field Guide to Rocky Mountain Wildflowers*. Boston: Houghton Mifflin.

Dana, Mrs. William Starr. [1893] 1963. *How to Know the Wild Flowers*, Rev. and enl. New York: Dover. Arranged by flower color. The conversational quality of the writing is appealing.

Darlington, Henry T., Ernst A. Bessey, and Clive R. Megee. 1945. *Some Important Michigan Weeds*. Special Bulletin 304. East Lansing, Mich.: Michigan State College Agricultural Experiment Station. Out of print, but watch for it, as it illustrates many of the common weeds of Michigan.

Dodge, Natt N. 1967. *100 Roadside Wildflowers of Southwest Uplands*. Globe, Ariz.: Southwest Monuments Association. Full-color illustrations of 100 flowers.

Fassett, Norman C. 1976. *Spring Flora of Wisconsin*. 4th ed. Rev. and enl. Madison, Wisc.: University of Wisconsin Press.

Hotchkiss, Neil. 1967. *Underwater and Floating-leaved Plants of the United States and Canada*. New York: Dover. Black-and-white drawings, short descriptions, and simple keys.

House, Homer D. 1974. *Wild Flowers*. New York: Macmillan. Reprint. Three hundred sixty-four full-color illustrations with complete descriptive text.

Niering, William, and Nancy Olmstead. 1979. *Audubon Society Field Guide to North American Wildflowers* (Eastern Region). New York: Alfred A. Knopf.

Peterson, Roger Tory, and Margaret McKenny. 1974. *Field Guide to Wildflowers of Northeastern and North-central North America*. Boston: Houghton Mifflin.

Rickett, Harold William. 1953. *Wild Flowers of America*. New York: Crown.

Smith, Helen V. 1966. *Michigan Wildflowers*. Bulletin 42. Bloomfield Hills, Mich.: Cranbrook Institute of Science. Well illustrated, with simple keys, this treats most of the more prominent wildflowers of Michigan.

Spellenberg, Richard. 1979. *Audubon Society Field Guide to North American Wildflowers* (Western Region). New York: Alfred A. Knopf.

Trelease, William. [1931] 1967. *Winter Botany*. Reprint. New York: Dover. A guide to identifying woody plants in winter.

U.S. Department of Agriculture. 1971. *Common Weeds of the United States*. New York: Dover. Almost any common weed you are likely to encounter can be identified from its pages.

Van Bruggen, Theodore. 1971. *Wildflowers of the Northern Plains and Black Hills*. Interior, S.D.: Badlands Natural History Association. Color photos of common wildflowers of this region, with comments on natural history and uses. Out of print.

Weber, William A. 1976. *Rocky Mountain Flora*. 5th ed. Boulder, Colo.: Colorado Associated University Press. Line drawings, keys, abbreviated descriptions included in the keys.

Wharton, Mary E., and Roger W. Barbour. 1971. *Guide to the Wildflowers and Ferns of Kentucky*. Lexington, Ky.: University Press of Kentucky. Very pleasing all-color illustrations of Kentucky's more prominent wild flowers and ferns, with brief accounts.

George Lawrence's *An Introduction to Plant Taxonomy* (Macmillan Company, 1955) is an excellent choice for a serious plant collector. It is packed with information about the theory and practice of plant taxonomy (the identification, classification, and naming of plants), with many pictures to help clarify each term used. For more advanced students, the longer version of the book (Lawrence, George H.M. 1951. *Taxonomy of Vascular Plants*. New York: Macmillan.) is an excellent choice.

Manuals for Plant Identification

Dean, Charles C. 1940. *Flora of Indiana*. Indianapolis, Ind.: Stechert. Out of print, but with luck you may find a copy in a used-bookstore.

Fernald, Merritt Lyndon. 1950. *Gray's Manual of Botany*. 8th ed. New York: American Book. Many people consider this the definitive manual for Northern and Eastern United States and Canada. Out of print, but available in libraries and, occasionally, in used condition.

Gleason, Henry A., and Arthur Cronquist. 1963. *Manual of Vascular Plants of Northeastern United States and Adjacent Canada*. Princeton, N.J.: D. Van Nostrand. No illustrations. The keys closely follow those of Britton and Brown (see next below).

Gleason, Henry A. 1963. *The New Britton and Brown Illustrated Flora of the Northeastern United States and Adjacent Canada*. 3 vols. New York: Hafner Press/Macmillan. Copiously illustrated, with comprehensive keys, this makes an excellent reference for amateur, student, and professional taxonomist. Usually referred to as Britton and Brown. Out of print, but still a standard reference. An older (1913) version of this work is available from Dover.

Muenscher, Walter Conrad. 1980. *Weeds*. 2d ed. Ithaca, N.Y.: Cornell University Press. Discusses the characteristics, habits, importance, dissemination, and control of the weeds of the northern United States and Canada, with emphasis on identification and control.

Radford, A., C.R. Bell, and H.E. Ahles. 1968. *Manual of the Vascular Flora of the Carolinas*. Chapel Hill, N.C.: University of North Carolina Press. Out of print.

Hickett, Harold William. 1966—. *Wild Flowers of the United States*. General editor William C. Steere. Collaborators Rogers McVaugh et al. New York: McGraw-

Hill. The six magnificently illustrated volumes of this series are of use to amateur and professional alike. The volumes are: 1. The Northeastern States. 2. The Southeastern States. 3. Texas. 4. The Southwestern States. 5. The Northwestern States. 6. The Central Mountains and Plains.

Small, John K. [1933.] 1972. *Manual of the Southeastern Flora.* Reprint. New York: Hafner Press/Macmillan. Out of print.

Steyermark, Julian A. 1963. *Flora of Missouri.* Ames, Iowa: Iowa State University Press.

Strasbaugh, P.D., and Earl L. Core. 1979. *Flora of West Virginia.* Morgantown, W. Va.: Seneca Books. Line drawings, keys, descriptions.

Voss, Edward G. 1972. *Michigan Flora,* Vol. 1. Bulletin 55. Ann Arbor, Mich.: Cranbrook Institute of Science, Bloomfield Hills, and the University of Michigan Herbarium. Vol. 1 of a projected 3-vol. work, this is an excellent, accurate, and highly usable guide to the gymnosperms and monocots of Michigan.

Special Categories

Aquatic Plants

Fassett, Norman C. 1966. *A Manual of Aquatic Plants.* Madison, Wisc.: The University of Wisconsin Press. With many black-and-white illustrations, and keys, this book is useful to both amateurs and professionals.

Voss, Edward G. 1967. A vegetative key to the genera of submersed and floating aquatic vascular plants of Michigan. *The Michigan Botanist* 6(2):35–50.

Trees and Other Woody Plants

Barnes, Burton V., and Warren H. Wagner, Jr. [1931] 1981. *Michigan Trees: A Guide to the Trees of Michigan and the Great Lakes.* Reprint. Ann Arbor, Mich.: University of Michigan Press. A highly recommended reissue of an older work by Charles Otis.

Berry, James Berthold. [1924] 1964. *Western Forest Trees*. Reprint. New York: Dover. A guide to the identification of trees and woods for students, teachers, farmers, and woodsmen.

Billington, Cecil. 1949. *Shrubs of Michigan*. 2d ed. Bulletin 20. Bloomfield Hills, Mich.: Cranbrook Institute of Science. Drawings, descriptions, distribution maps of Michigan's shrubs and vines.

Harlow, William M. [1946] 1959. *Fruit Key and Twig Key to Trees and Shrubs; Fruit Key to Northeastern Trees*. Reprint. New York: Dover.

Harlow, William M. 1957. *Trees of the Eastern and Central United States and Canada*. New York: Dover. Keys, descriptions, photographs of leaves, buds, fruit, and bark.

Harrar, Ellwood S., and J. George Harrar. 1962. *Guide to Southern Trees*. 2d ed. New York: Dover Publications. Keys, drawings of twigs and fruits, and descriptions of southern trees westward through Texas.

Hosie, R.C. 1973. *Native Trees of Canada*. 7th ed. Canadian Forestry Service, Department of the Environment. Difficult to find but highly recommended.

Keeler, Harriet L. [1903] 1969. *Our Northern Shrubs and How to Identify Them*. Reprint, updated. New York: Dover. Illustrated by photographs of twigs and flowers. The descriptions incorporate bits of history and uses of the plants.

Little, Elbert L. 1980. *The Audubon Society Field Guide to North American Trees* (Eastern Edition). New York: Alfred A. Knopf.

Muenscher, W.C. 1950. *Keys to Woody Plants*, 6th ed., rev. Ithaca, N.Y.: Cornell University Press. Nonillustrated keys to woody plants in both summer and winter condition.

Petrides, George A. 1972. *A Field Guide to Trees and Shrubs*. 2d ed. Boston: Houghton Mifflin. Many semidiagrammatic drawings, descriptions, and both winter and summer keys.

Sargent, Charles Sprague. [1922] 1961. *Manual of the Trees of North America*. 2 vols. Reprint. New York: Dover.

Symonds, George W.D. 1973. *The Tree Identification Book*. New York: William Morrow.

Symons, George W.D. 1963. *The Shrub Identification Book*. New York: M. Barrows.

Wharton, Mary E., and Roger W. Barbour. 1973. *Trees and Shrubs of Kentucky*. Lexington, Ky.: University Press of Kentucky. Copiously illustrated in black-and-white and color.

Grasses

Although grasses are included in many of the manuals listed above under the topic "Manuals for Plant Identification," they do make up one of the more difficult groups, and merit individual treatment. Here are four especially helpful books.

Chase, Agnes. 1977. *First Book of Grasses*. 3d ed. Washington, D.C.: The Smithsonian Institution. The structure of grasses explained for beginners. This book will guide you through the first understanding of the specialized terminology and structure.

Hitchcock, A.S. 1971. *Manual of the Grasses of the United States*. 2d ed. Revised by Agnes Chase. U.S.D.A. Misc. Publication No. 200. Washington, D.C.: U.S. Government Printing Office. Keys, many line drawings, descriptions, distribution, plus a listing of synonyms of all the grasses of the United States.

Pohl, Richard W. 1978. *How to Know the Grasses*, 3d ed. Dubuque, Iowa: Wm. C. Brown. Picture keys, helpful tips on collection and study.

Voss, Edward G. 1972. *Michigan Flora*. See above under "Manuals for Plant Identification."

Orchids

Beautiful, often rare, all of these plants are on the threatened or endangered species list. Study them in the field, but do not collect or attempt to transplant them. They demand specialized growing conditions, and almost certainly would not survive in a garden.

Case, Frederick W., Jr. 1964. *Orchids of the Western Great Lakes Region*. Bulletin 48. Bloomfield Hills, Mich.: Cranbrook Institute of Science. Beautifully illustrated in black-and-white and color, this book will help the student learn to identify these increasingly rare plant treasures. Out of print, but soon to be reissued, all in color.

Luer, Carlyle A. 1972. *The Native Orchids of Florida*. Bronx, N.Y.: New York Botanical Garden.

Luer, Carlyle A. 1972. *The Native Orchids of the United States and Canada Excluding Florida*. Bronx, N.Y.: New York Botanical Garden.

Cultivated Plants

Most of the plant identification books deal only with wild plants. The books listed below will give some guidance to the scientific names of cultivated plants.

Bailey, Liberty Hyde. 1943. *The Cyclopedia of Horticulture*. 6 vols. New York: Macmillan. Illustrated with colored plates, 4,000 engravings in the text, and 96 full-page cuts. For the amateur and professional, a magnificent resource.
Bailey, Liberty Hyde. 1949. *Manual of Cultivated Plants*. Rev. ed. New York: Macmillan. Keys, scientific descriptions, some drawings. From ferns to composites, a 1-volume guide to thousands of cultivated plants, including trees and shrubs.
Bailey, Liberty Hyde, and E.Z. Bailey. 1976. *Hortus Third. A Concise Dictionary of Plants Cultivated in the United States and Canada*. New York: Macmillan.
Everett, Thomas H., Ed. 1967. *Reader's Digest Complete Book of the Garden*. Pleasantville, N.Y.: Reader's Digest Association. This surprisingly helpful book has enabled me to find the names and facts of cultivation of many house and garden plants, and is rich in odd bits of information not found elsewhere. Out of print.
Everett, Thomas H. 1980. *The New York Botanical Garden Illustrated Encyclopedia of Horticulture*. 10 vols. New York: Garland Publishing. An immense amount of information—detailed descriptions of plants and plant families, practical directions for culture outdoors and in—provides answers to many questions.

Cryptogams

Plants that do not have flowers or seeds, but propagate themselves by means of spores, are called "cryptogams," which means "hidden marriage." Among these are the algae, fungi, mosses, liverworts, ferns, and their allies.

Algae

The algae form such a specialized group that a course of study is required in order to identify them. There are both freshwater and marine forms. Some books that will help with identification of the larger marine algae of both the Atlantic and Pacific coasts are:

Collins, F.S. [1905–1909] 1970. *The Green Algae of North America* and Supplements 1 and 2. Reprint. Monticello, N.Y.: Lubrecht and Cramer.

Dawson, E. Yale. 1956. *How to Know the Seaweeds*. Dubuque, Iowa: Wm. C. Brown. Out of print, but possibly to be found in used-bookstores and in libraries.

Guberlet, Muriel Lewin. 1956. *Seaweeds at Ebb Tide*. Seattle, Wash.: University of Washington Press. Drawings and descriptions to help you name the larger seaweeds. Out of print.

Setchell, W.A., and N.L. Gardner. [1920–1925] 1968. *Algae of Northwestern America*. Reprint. Monticello, N.Y.: Lubrecht and Cramer. A classic work.

Smith, G.M. 1944. *Marine Algae of the Monterey Peninsula*. Stanford, Calif.: Stanford University Press. Out of print.

Taylor, William R. 1957. *Marine Algae of the Northeastern Coast of North America*. Ann Arbor, Mich.: University of Michigan Press. Suitable for the western Atlantic from Virginia to the Arctic ocean.

Fungi

For other than the more common fungi, you will need the direction of an instructor. Five guides to the mushrooms are listed below.

Krieger, Louis C.C. [1936] 1967. *The Mushroom Handbook*. New York: Dover. Reprinted with new information. Keys, descriptions, Louis Krieger's lovely paintings, even recipes.

Smith, Alexander H., et al. 1981. *How to Know the Non-gilled Mushrooms*. 2d ed. Pictured Key Nature Series. Dubuque, Iowa: Wm. C. Brown.

Smith, Alexander H., and Nancy Weber. 1980. *The Mushroom Hunter's Field Guide*. Rev. and enl. Ann Arbor, Mich.: The University of Michigan Press. Copiously illustrated in color, with identification marks, edibility, and when and where to find it notes on each species.

Smith, Helen V., et al. 1979. *How to Know the Gilled Mushrooms.* Pictured Key Nature Series. Dubuque, Iowa: Wm. C. Brown.

Stevens, Russell B., Ed. 1981. *Mycology Guidebook.* Rev. ed. Seattle, Wash.: Mycological Society of America, University of Washington Press. The standard reference book for mycological techniques.

Lichens

Hale, Mason E. 1979. *How to Know the Lichens.* 2d ed. Pictured Key Nature Series. Dubuqe, Iowa: Wm. C. Brown Co.

Mosses and Liverworts

Conard, Henry S., and Paul L. Redfearn, Jr. 1979. *How to Know the Mosses and Liverworts.* 2d ed. Pictured Key Nature Series. Dubuque, Iowa: Wm. C. Brown Co. Good introductory material on how to collect and study these plants.

Crum, Howard A. 1983. *Mosses of the Great Lakes Forest.* 3d ed. Ann Arbor, Mich.: University Herbarium, The University of Michigan. For the serious student and taxonomist.

Crum, Howard A., and Lewis E. Anderson. 1981. *Mosses of Eastern North America.* New York: Columbia University Press. For advanced workers.

Darlington, Henry T. 1964. *The Mosses of Michigan.* Bulletin 47. Bloomfield Hills, Mich.: Cranbrook Institute of Science. Keys, short descriptions, many illustrations. For the serious student.

Grout, A.J. 1924. *Mosses with a Hand-lens.* 3d ed. Newfane, Vt.: Published by the author. Simple keys to mosses and liverworts of the northern and middle Atlantic states. Out of print, but look for it.

Grout, A.J. [1903] 1972. *Mosses with Hand-lens and Microscope.* Reprint. Augusta, W. Va.: Eric Lundberg. Abundant illustrations and excellent text. Mosses of the northeastern United States. Reprint of an old work.

Grout, A.J., Ed. 1972. *Moss Flora of North America.* 3 vols. New York: Hafner Press/Macmillan. Descriptions of all species of mosses of the United States including Alaska; Canada, Newfoundland, and Greenland, with many illustrations. Out of print, but available in libraries.

Steere, William Campbell. 1940. *Liverworts of Southern Michigan.* Bulletin 17. Bloomfield Hills, Mich.: Cranbrook Institute of Science. Keys, descriptions, photographs.

Ferns and Their Allies

Billington, Cecil. 1952. *Ferns of Michigan*. Bulletin 32. Bloomfield Hills, Mich.: Cranbrook Institute of Science. Keys, many line drawings, distribution maps, descriptions, plus considerable other material on structure, methods of study, cultivation, folklore, etc. Very usable for Michigan species.

Cobb, Boughton. 1977. *A Field Guide to the Ferns and Their Related Families of Northeastern and Central North America*. Boston: Houghton Mifflin.

Foster, F. Gordon. 1971. *Ferns to Know and Grow*. New York: Hawthorn Books. Formerly *The Gardener's Fern Book*, 1964. Princeton, N.J.: D. Van Nostrand. A guide for the gardener; a reference for the nature-lover. Many drawings and photos by the author. Bridges the gap between native and cultivated ferns. Out of print, but you may be able to find it at book sales.

Parsons, Frances Theodora. [1899] 1961. *How to Know the Ferns*. 2nd ed. Reprint. New York: Dover. This reprint of a charming older work is best described in the words of its own subtitle, "A guide to the names, haunts, and habits of our common ferns." The discursive style of writing adds to its flavor. Many illustrations.

Wherry, Edgar T. 1961. *The Fern Guide. Northeastern and Midland United States and Adjacent Canada*. Philadelphia: Morris Arboretum. The best fern field guide. Out of print, but useful if you can find it.

NOTE: If I have failed to comment on any of the books listed above, it implies only that I am not acquainted personally with them, and that they were recommended to me by others.

Plant identification can be a lifelong avocation, partly because it is not easy. A few plants are unmistakably themselves and no other, but most, however, tend to offer problems. In the throes of the first struggles with identification, it may seem that the manual is wrong, or that an unusual hybrid is at hand (either of which conclusions *may* be the truth). There is an art as well as a science to the weighing of the different factors, the deciding which variations may be individual, or which denote a separate species. Only practice will help. Be encouraged. Like the study of music, the study of plants is a continuum, from the stumbling first steps of the beginner to the elegant work of the expert, and *no one* knows every plant in the world. Just enjoy the process of learning.

4

Labeling

The label is of great importance to any scientifically valuable specimen. Without a label there is only a dead plant; with it is a contribution to scientific records—possibly significant. Amateurs have been known to bring in some very interesting material.

Even if you do not expect your specimens to be of value to science, they may someday take on such a value if they are carefully preserved and labeled adequately. Old amateur collections sometimes may contain the only representative specimens from a particular place.

Make the labels before attempting to mount or package a collection. Labels should be made of 50 to 100 percent rag content bond paper and they must be typewritten, printed, or written in permanent black ink. Assume that your work will be of value hundreds of years later. Faded ink and incomplete or illegible entries can cause later researchers to tear their hair.

A good label size is 4¼ × 2¾ inches (11 × 7 cm), large enough to contain complete data, yet small enough not to take up too much room on the herbarium sheet. However, a 3- × 5-inch label has certain advantages. Because it is the same size as a standard file card, I use this

```
┌─────────────────────────────────────────────────────────────┐
│                      Title of Label                          │
│                                                               │
│          PLANTS OF (some geographical region)                 │
│                                                               │
│   Genus specific epithet Author                               │
│   Family                                                      │
│   Common name                                                 │
│   Locality, as exactly as possible                            │
│                                                               │
│   Township, range & section    or    lat. & long.            │
│   Habitat                                                      │
│   Miscellaneous observations                                  │
│                                                               │
│   Collector's name                         Coll. No.          │
│   Identifier's name                        Date               │
│                                                               │
└─────────────────────────────────────────────────────────────┘
```

```
┌─────────────────────────────────────────────────────────────┐
│         Eastern Michigan University Herbarium (EMC)           │
│                     Ypsilanti, Michigan                       │
│         PLANTS OF U.S.A., MICHIGAN, WASHTENAW COUNTY           │
│                                                               │
│   Rhus typhina L.                                             │
│   Anacardiaceae                                               │
│   Staghorn Sumac; Velvet Sumac                                │
│   Pittsfield Twp., Thomas & Morgan Roads, SE corner.          │
│   T3S, R6E, NE 1/4 Sect. 22.                                  │
│   Roadside, poor dry clay soil, full sun.  Branch of          │
│   10-ft. shrub; flowers yellow-green; sap milky.  With        │
│   Virginia creeper, wild carrot.                              │
│                                                               │
│   Coll.  Ruth B. (Alford) MacFarlane    Coll. No. 2133        │
│   Id.    RBM                             25 VI 1975           │
│                                                               │
└─────────────────────────────────────────────────────────────┘
```

Sample plant labels: top—in diagrammatic form; bottom—with information filled in

size and make carbon copies at the same time, with which I maintain a catalog of my plant collection. Other label sizes must be used for specimens stored in vials or boxes of smaller size.

Two labels are pictured on this page, one in diagrammatic form, the other filled in as an example.

If many labels are needed, have basic label forms preprinted. For the relatively small quantities used by most beginners, many people now type a sheet of eight label forms on standard typing paper of high quality.

This sheet is photocopied on 50 percent or 100 percent rag paper, and the labels are cut apart, completed, and glued onto the herbarium sheets. These photocopies are considerably cheaper than printed labels and can look clean and professional if done on a good copier. Do not use carbon copies for labels, since these will smudge and appear carelessly done.

Each label has a title, which may be the name of the herbarium where the collection is housed (EASTERN MICHIGAN UNIVERSITY HERBARIUM [EMC], Ypsilanti, Michigan); the name of a special collection (MEXICO, State of Jalisco or ALGAE OF THE PACIFIC COAST); or the name of an individual (HERBARIUM OF ROBERT PETER, M.D.). Some collectors include a small map of the collecting area on the label and place a colored dot at the approximate collecting site.

Someone consulting a specimen later will be first interested in the name of the plant and its source, so these should be readily apparent. Other information to be included is that which is not evident from the plant itself in its preserved condition.

A label should contain:

- Genus, specific epithet, and author of the species name
- Family
- Common name
- Locality
- Habitat notes
- Miscellaneous observations
- Collector's full name
- Collection number
- Date
- Name of the person identifying the plant

Genus, Specific Epithet, Author, and Family

The name of a particular species (for example, *Rosa carolina*) consists of two parts, called the genus (*Rosa*) and the specific epithet (*carolina*). This two-part name, or binomial, as well as that of the family (Rosaceae), will be in Latin, as scientific names are, and may be obtained from the plant identification manual. The genus is always capitalized, the specific epithet usually not; both are italicized or underlined. Following the species name is a name, an initial, or an abbreviation. This represents the name of the person(s), called the author(s), who named the plant. Copy this down exactly as it appears, perhaps simply "L." (for Linnaeus) or "(L.) Michx." (first named by Linnaeus, reclassified by Michaux). This is part of the

Rosa carolina

scientific name (*Rosa carolina* L.). The author is *not* italicized. Family names need not be italicized either.

Common Name

Common names may be included as a matter of interest. They can be very useful and should be given if possible. Regional differences in common names occur, and often give clues to the uses of the plants.

Locality

Include the country, state or province, county or parish, township or city, and the exact location, which may be a street address or be indicated by the number of miles it is distant from a given point, or by some other firm reference. The description should guide someone to the same spot 50 years hence. The locality data is recorded from large units to small ones, as they are listed previously. This format will aid a later researcher sorting through specimens for those of a particular geographic region, perhaps "U.S.A., Michigan, Wayne County." Sometimes this part of the data will be placed at the top of the label, directly beneath the name of the herbarium or collection, and need not be repeated.

Although political boundaries such as counties and townships change slowly, they do change, and lesser landmarks may appear or disappear. Roads are built or abandoned; hills are leveled or swamps filled in; cities grow old or shrink. A well-defined present-day site may become unrecognizable in 50 years.

An ingenious system, the United States Land Survey system, was devised by Thomas Jefferson and adopted by Congress in 1785. It is a grid of imaginary lines, laid out in squares 6 miles on a side, called Congressional townships. These may or may not coincide with named townships. Congressional townships in each state or territory are counted off in four directions from the point where a true east–west line, called a Base Line, intersects a true north-south line, called a Principal Meridian. These lines correspond to latitude and longitude (see the illustration on page 57).

Land in the first tier of townships north of the Base Line is designated as being "Township One North" (T.1 N.); land in the second tier is T.2 N., etc. South of the Base Line, land is designated as T.1 S., T.2 S., etc. That takes care of numbering from north to south. The townships marked off east and west of the Principal Meridian are called Ranges. Those east of the Principal Meridian are "Range One East" (R.1 E.), "Range Two East" (R.2 E.), etc. and those west are R.1 W., R.2 W., etc. Because of the Earth's spherical shape, the meridians converge toward the poles. To compensate for this, at the north side of every set of four townships, a correction is made, setting the townships back to a 6-mile width.

Each 6-mile-square township is divided into 36 sections, numbered as a surveyor would walk back and forth across the township (as indeed he did), beginning at the northeast corner. Each section, in common usage, is divided into square quarters, and those again into quarters that comprise 40 acres (16.19 hectares) each, that being a unit in which land is commonly sold. Hence comes the term the *back forty*. When describing a piece of land, you can do so briefly and accurately by saying, "the SE 1/4 of the SW 1/4 of Sec. 33, T.2 N., R.2 E." NOTE: On plant labels, to save space, you might abbreviate to T2N, R2E.

Because this grid exists only in the imagination, it never changes and remains an accurate method of describing locality. Township, Range, and Section descriptions (often referred to as TRS) can be found on topographic maps, on some county maps and those of special areas such as a national forest, and in plat books. Topographic maps also show land altitudes, building and road locations, and forested areas. Public libraries often maintain map collections, or maps may be obtained from agencies like auto clubs, state departments of natural resources, county offices, or sporting goods stores. Otherwise, write to:

> United States Geological Survey
> Map Distribution
> 1200 Eads Street
> Arlington, Va., 22202

for their list of maps and prices.

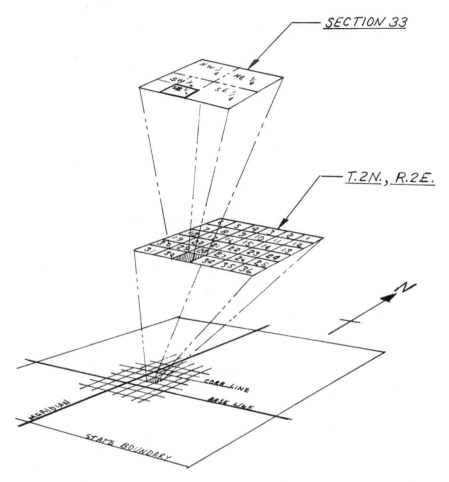

The Congressional-township grid system (drawn by Richard MacFarlane)

Serious collectors should always look up the TRS and include it in their label information.

Not all areas are laid out under this grid system, but it applies to all of the United States except the 13 original states, some of the eastern states, and Texas. In Kentucky, for example, topographic maps will be marked off in the degrees, minutes, and seconds of latitude and longitude, in which case, include these on the label.

Habitat Notes

Please refer to the discussion of habitat notes in Chapter 1.

Miscellaneous Observations

Material to be included under miscellaneous observations is described in Chapter 1. This information should always apply to the particular plant here collected, and not be general remarks about the species. An exception might be notes on the usefulness of the plant. If the local residents use the plant as a dye, or as a food, that is worth recording.

Collector's Full Name

Include the collector's first and last names and at least a middle initial. If labels are typed, it is a personal touch to write the collector's name. Who knows, you may turn out to be one of the "greats" of plant study. If your writing is not easily legible, type or print your name, too.

Collection Number

The collection number should appear on the label. Please see Chapter 1 for a discussion of collection numbers.

Date

The date should be complete—day, month, and year—with the month either in Roman numerals or abbreviated (11 XII 1977 or 11 Dec 1977). If Arabic numerals are used for both day and month, someone reading the label years later, and finding "7/9/1983," will not be sure if the plant was collected on the seventh day of September or the ninth day of July.

Name of the Person Identifying the Plant

Often someone more expert than the collector will actually identify the plant. This person's name should appear on the label after "Id." (identifier) or after "*Fide*" (corroborator).

Don't throw away a good specimen, with complete data, because it cannot be identified. Put all available information on the label, leaving room for the plant name, and hope an expert will identify it.

Labels for Boxes, Jars, and Vials

Labels for specimens in boxes are usually the same as for those on herbarium sheets. One is placed outside on the lid or side of the box. One is placed inside. For boxes smaller than usual label size, make smaller labels. Folding labels is poor practice.

If the plants, whether algae, fungi, or fruits, are preserved in liquid, use as small a label as possible, made of 100 percent rag bond paper so it won't disintegrate, and print on it in India ink or waterproof technical drawing ink. To make my printing small enough, I use either a crow quill pen or a technical drawing pen with an 00 or 000 point, using a magnifier as I print. All the data described above should be included. It is possible to identify collections in vials by only a collection number or a catalog number, but there is always great danger that the field notebook or the catalog will become separated from the vials and lost, after which the specimens become valueless. The only really safe way is to get that label into the vial or jar.

While specimens are "in process," that is, somewhere between the time of collection and the printing of the label, temporary labels in pencil on 100 percent rag bond paper will keep continuity. Sometimes the in-process period can last years.

It is easy to set up a file or cross-file of specimens as labels are typed or written. (Use the larger labels such as those for herbarium sheets.) Either make a carbon of your label, or photocopy the labels. In addition to a card for the specimen, it is useful to have one by locality (as Washtenaw County, Ontonagon County, etc.). In a herbarium, further files by collector or by accession number may be kept.

Neatness and accuracy in spelling are essential when labeling a collection. Misspelled words may make later workers question the validity of the rest of the work.

5

Mounting and Packaging

Having collected, pressed or otherwise preserved plant specimens, identified them, and made labels, the next concern will be with mounting or packaging the material. I am going to describe the methods for preparing plants for scientific collections. These instructions may be adapted for amateur collections, too. Suggestions for the ornamental uses of preserved plants will be found in Chapters 9 and 10.

Herbarium Sheets

Pressed plants to be used for scientific purposes are mounted on herbarium sheets of standard size. There are several methods of mounting plants. Choose among these methods and practice with scrap materials until skill is developed, so as not to spoil good specimens.

Glass-Plate Method

The materials needed are listed in the following. They can be obtained from biological supply houses and special herbarium supply companies. The less specialized materials, the brushes, cheesecloth, and waxed paper, may be purchased locally.

- *Herbarium paper*—11½ × 16½ inches (29 × 42 cm), of good quality rag content, weight about 45M to 68M. You may use the less expensive 25–50 percent rag paper, but 100 percent rag content is best for professional collections. It lasts longer and resists yellowing.

- *Glue; Brown glue*—called "hide" glue, a by-product of packing-house waste, is commonly used for most plants. *White glue*, of the Elmer's glue type, is used for twigs, waxy leaves, and other plant forms more difficult to mount. It is also called casein glue. Some herbarium workers use white glue exclusively even with the glass-plate method.

- *Glass or plastic plate*—12 × 18 inches (30 × 46 cm) for spreading out the glue, when a number of plants are to be mounted. Use plate glass with ground edges, or plastic of similar thickness, about ⅛ to ¼ inch (3 to 6 mm) thick.

- *Brushes*—1 inch wide (2.5 cm), with soft bristles.

- *Waxed paper*—12 × 18-inch (30 × 46 cm) sheets.

- *Blotters*—12 × 18 inches (30 × 46 cm), or folded newspapers.

- *Pads of polyurethane foam*—12 × 18 inches (30 × 46 cm), or pads of folded newspapers, or sandbags.

- *Corrugated cardboard ventilators (corrugates)*—12 × 18 inches (30 × 46 cm), double faced, with each side smooth.

- *Labels*—See the chapter on labeling for types of labels and for the data to be written on each one.

- *Library paste*—for affixing labels and packets. (Or you can use white glue.)

- *Packets or fragment folders*—See pages 70–72.

- *Dish of water*

- *Forceps*

- *Cheesecloth square*

Procedure for Mounting

Arrange materials in a convenient place, and always follow each of the steps in the same order, so that none will be omitted. I usually work in a U-shaped space. On my left are pressed plants still in papers, with labels already typed. Before me are a pile of herbarium paper, the sheet of

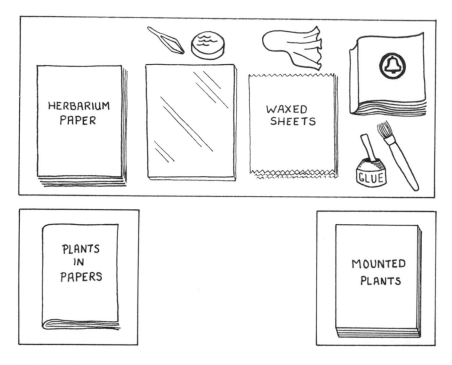

Mounting setup

paper on which to mount the first plant, the glass or plastic plate for gluing, the glue containers, a dish of water, a dampened cheesecloth square, sheets of waxed paper, and packets. On my right are the library paste for labels and an old telephone directory as a pasting pad. Farthest to the right will be the pile of mounted plants. Close at hand are brushes and forceps, and piles of blotters and pads and corrugated cardboard ventilators, also called corrugates.

Begin the pile of mounted plants with a piece of corrugated cardboard. Unless only a few plants are being mounted, it is best to label the pile appropriately ("Collection numbers 2316—2340," or "Plants from Colorado trip") with a paper strip that projects out the end. Lay a blotter on the cardboard to absorb moisture from the glue. Folded newspapers, several sheets thick, make good blotters.

Measure each plant on the sheet first. It should fit without projecting over the edge at any spot and without obscuring any part of the label, leaving room somewhere on the sheet for a small packet or fragment folder. (See the illustration on page 72.) Place the plant to best display important features, such as the flowers. Either end of the plant may lie toward the top of the sheet. Try to leave some space above the label for later annotations.

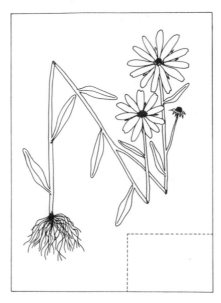

Mounting a large plant, N-fold

First, paste the label in the lower right-hand corner of the herbarium sheet. An old telephone directory makes a pasting pad; discard its pages as they become soiled. Keep the label edges even with those of the sheet and avoid wrinkles. Use a smooth stick like a tongue-depressor to smooth out the label, or cover it with a piece of paper while smoothing it with your fingers—anything to avoid smearing the ink. It looks best if a margin of 1/16 inch (1 or 2 mm) is left between the label and the paper edge. At this stage, and at every other, neatness is important if a professional looking finished product is to be made. A crooked label will be a continuing reproach.

Spread brown glue and some water on the glass or plastic plate, mixing and spreading it with the brush to obtain the right consistency. Only practice can show what is "right" for any given plant; not so thin the plant doesn't adhere, or so thick that there is too much on the sheet. It is important not to get too much glue on the flowers, just enough to fasten them down. A later researcher may need to remove a flower for dissection and study. Too much glue would obscure important features. If many flowers are on the specimen, try to mount the specimen so that some flowers receive *no* glue, or else remove a few to be put in a separate packet on the sheet.

Remembering the plant's placement on the sheet, drop it into the glue on the plate, and gently tap down any parts that do not quite touch until every surface is coated. Lift the plant from the glue with forceps and fingers and place it on the herbarium sheet. Someone with a steady hand can drop it on, which helps to spread the leaves; otherwise, inch down. Blot up any excess glue with the dampened cheesecloth square.

For woody twigs, thick stems, or waxy leaves, use white glue applied with a brush or directly from a dispenser, for extra adhesive strength.

Set the herbarium sheet on the corrugated cardboard and blotter at the right and cover it with a piece of waxed paper. Over the waxed paper set a blotter and a thick pad of newspaper, or a rectangle of flexible plastic or rubber foam, whichever is more appropriate to the thickness or unevenness of the particular plant. Mrs. Jennie Dieterle, of the University of Michigan Herbarium, adapted a method used at Kew Herbarium, England. She designed sandbags of closely woven unbleached muslin, 12 X 18 inches (30 X 46 cm), each one filled with 12 pounds (5.4 kg) of mined silica sand, which resemble flattish pillows 1½ inch (4 cm) thick. They make excellent weights for glued plants, fitting every contour and holding the plants firmly in place. If the plants are of uneven thickness and neither foam pads nor sandbags are available, folded pads of newspaper can help to equalize pressure and press the plant firmly to the sheet as it dries.

Place weights on the pile of plants and allow them to dry for 24 hours.

As work proceeds, keep the glass plate clean by scraping away any plant fragments or dirt. You do not want the pollen of a ragweed, for example, to appear on the next plant mounted. Wash and dry the glass plate and all other tools at the end of each session of plant mounting.

Right after the plant has been glued, and before adding it to the pile, glue a fragment folder or packet to it. Especially in professional collections, every scrap of the plant may be precious, and pieces do fall off. Also, in working with pressed specimens, a researcher who has removed a flower for study should save it. I try to press extra flowers, or remove some that would be hidden beneath the mounted plant. All these bits and pieces can be placed in a packet. A useful small-sized packet can be made from one-half or one-quarter of a sheet of standard sized typing paper, of 100 percent rag bond, folded as described for the larger packet (see pages 70–72). Mount it so the flaps can be turned under, putting only a small circle of glue in the center of the packet to leave room for the turning under. Packets can go anywhere on the sheet, preferably along the margin. It is a good idea to place them in different places from sheet to sheet to help balance specimen thickness on the storage shelves. Remember to keep the space directly above the label clear for annotation labels.

Sometimes larger packet sizes are called for. Be guided by the problem at hand. Excessively brittle plants, such as *Chara*, or plants too small to be glued directly to the sheet, may be placed in packets and the whole packet glued on.

Purchase large quantities of packets or fragment folders from a herbarium supply house to save time.

Painting-with-Glue Method

Instead of spreading glue on a glass plate and dropping the plants on it, turn over each plant and paint the back of it with glue. White glue works best for this method. Some of the plants may have thick twigs or waxy leaves, and white glue is always best for this type. Thin the glue with water if necessary to the consistency of thick cream, and apply it carefully with a medium-soft-bristled brush about 1-inch wide (2.5 cm). For twigs only, apply the glue directly from its dispenser, using a steady hand.

Some plant mounters use this method for all plants. Particular care must be taken not to get white glue into the flowers, since it is harder to soften with water, once set, than brown glue is.

After the reverse side of the plant has been painted, carefully pick it up and turn it over and place it on the herbarium sheet. Blot away any excess glue with the dampened cheesecloth square.

Plastic Adhesives

Another method of gluing used in many herbaria is the use of dots or strips of plastic adhesive. Sold under the name of Archer's Adhesive by biological supply houses, the adhesive consists of ethyl cellulose and Dow Resin 276 dissolved in a basic formulation of 4 parts of toluene to 1 part of methanol. (The first two compounds are obtainable from the Dow Chemical Company, who will give you the exact formula for the adhesive, and the toluene and the methanol from chemical or biological supply houses.)

Lay the plant on the herbarium sheet, weight it down here and there with small weights, and apply dots or strips of the plastic adhesive. As it dries, the adhesive will strap the plant down. Spread the sheets over tables as they dry or pile them up by placing spacers between each sheet and the sheet of corrugated cardboard above it. Spacers can be little blocks of wood or pieces of dowel. When mounted this way, the specimens can be cut off the sheet if necessary, which may or may not be an advantage, depending upon their use.

Spray Gluing

A technique practical only in a large herbarium but worthy of description here is one now used at the University of Michigan Herbarium at Ann Arbor. Drs. William Anderson and Anton Reznicek have developed a spray-gluing method. Using a paint-spraying apparatus, water-soluble white glue, and working under a ventilating hood, two skilled technicians can mount up to 800 plants a day.

The two technicians stand at the spraying booth, which should be of comfortable working height. The first person positions the plant on a sheet of herbarium paper onto which the label has already been glued. If possible, some flowers are broken off and placed in a fragment folder to protect them from glue. Technician 1 turns over the plant, sprays it with

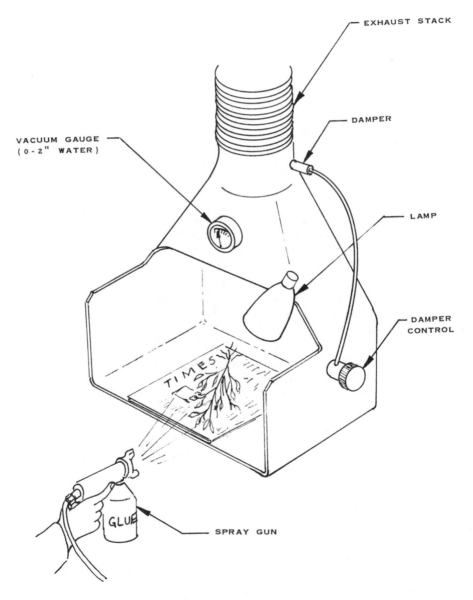

Spray-glue setup (drawn by Richard MacFarlane)

glue, picks it up with forceps, and places it right side up on the herbarium sheet, then removes any excess glue with a damp cloth. Technician 2 covers the sheet with waxed paper and places it in a pile thus: first, a corrugated cardboard, next a blotter, then the mounted specimen covered with waxed paper. Depending upon the thickness of the specimen, it is padded with a second blotter, a piece of foam plastic, or a sandbag. Another corrugated cardboard forms a base for the next pile of blotter, specimen, and pad.

When dry, the specimens may be reinforced with strips of cloth tape or by sewing.

The glue used in the sprayer is diluted RESYN 356262, obtainable from:

National Starch and Chemical Corporation
3641 S. Washtenaw Avenue
Chicago, Ill. 60632

Full-strength glue is always kept nearby for hand brushing onto bulky, waxy leaves.

The spray technique works especially well on filmy plants or those that are old and brittle, but there are some problems. There is less control over where the glue goes than with the hand-brushing method. There is no way to keep the glue out of flowers except, as mentioned above, by removing some blossoms and storing them in packets attached to the sheet, or by masking them. The spray may collect on fine hairs, giving them a beaded appearance resembling glandular hairs. Excess glue may get on the upper side of the plant.

A skilled operator can minimize these difficulties by masking certain parts of the plant while spraying, by using different dilutions of glue, and by working carefully, still producing a better result than the work done by unskilled workers. The greatest advantages are the increased speed of mounting and the gentler treatment of brittle specimens.

The components of the spraying system are relatively simple. The most expensive part is the hood and ventilating system, which contains an exhaust fan and a filter. The filter must be changed frequently. A light is installed to shine directly down upon the work in progress.

The glue-spraying part is composed of an air compressor (in this case a Binks Diaphragm Compressor, Model 34-2025) and a hand-held, trigger-operated spray gun. This gun must be cleaned regularly or it will clog up with glue.

Using this system, after initial practice, a careful operator and an assistant can rapidly reduce an accumulated backlog of plant specimens, and keep abreast of large numbers of acquisitions.

Spray Gluing, Using a Nylon-Thread Grid

A spray-glue method, adaptable for use in smaller herbaria, is used at the University of California, Santa Barbara, by Dr. Wayne R. Ferren, Jr.

A wooden frame about 12 × 18 inches (28 × 46 cm), slightly larger than a sheet of herbarium paper, is strung with a ½-inch (1.3 cm) grid of nylon thread.

The technician arranges the specimen on the herbarium sheet and places the packet and label on it (already glued or paperclipped for later gluing). A piece of sturdy corrugated cardboard is usually placed beneath the paper to facilitate movement of the material. The wooden frame is placed on top of the sheet and the whole thing is turned over. The cardboard and sheet are lifted off, leaving the plant on the grid. The specimen is then sprayed with a 1-to-1 or 1-to-2 solution of Elmer's glue and water, using a plastic spray bottle. Any extra glue sprays right on past the plant onto a paper beneath the grid. If any woody or coarse parts are present, 100 percent Elmer's glue is applied directly to them since the weaker solution does not hold these parts. One must *be careful not to move the plant on the grid,* or it will not return to its original position on the sheet. The herbarium sheet is laid back on to the glued specimen, with the wooden frame serving as a template for accurate placement, and the whole assembly turned back over to its original position. The grid is lifted off. Waxed paper is put over the specimen, followed by a blotter and a sandbag. A pile of sheets is made, with a corrugated cardboard placed between every few sheets and blotters, to equalize the weight.

This method is especially useful for aquatic plants or grasses whose leaves are limp and difficult to move once they are coated with glue.

Roller Method

A second method being tried at the University of California at Santa Barbara replaces the spraying step with the use of a small paint roller rolled into a paint well filled with Elmer's glue, then rolled onto the underside of the specimen. The roller is about 3 inches (8 cm) wide and is available at many hardware stores. The glue is diluted 1-to-1 or 1-to-2 with water. This method is not effective for grasses, aquatics, and other delicate plants.

Cloth Tape for Mounting and Reinforcement

An old-fashioned mounting method is to fasten the plant down with narrow strips of gummed cloth tape, using no glue at all. This has the advantage of leaving the specimen easy to remount, should that become necessary later, but it has the disadvantage of the specimen's being less firmly affixed.

Cloth tape is often used in addition to glue, to reinforce the mounting. The tape commonly sold in department stores or dime stores these days is plastic and is not suitable. Good cloth tape may be bought from herbarium supply houses. Cut it carefully into narrow strips of a suitable width for the plant material being reinforced, about ⅛- to ¼-inch wide (3 to 6 mm). Cut it into pieces, making the end cuts square. For a ⅛-inch (3 mm) diameter twig, a piece 1 inch (2.5 cm) long is about right. Using forceps, tuck the strip of tape around the twig and attach the ends to the paper, making it lie at right angles to the twig. Attention to such small details helps to make a neatly mounted specimen. Use a reasonable number of tapes, placed at points of strain to keep the twig from springing away from the paper, but not so many as to give a bandaged look.

Never use cellophane or other plastic tape. It discolors in time and ruins the specimen, besides becoming sticky when exposed to necessary fumigants.

Reinforcement by Sewing

A neat method of reinforcing a glued plant is by placing a few stitches here and there at points of strain. Use waxed white linen button thread. Prick two holes next to the stem or twig to be reinforced. On the back of the sheet, cover the two holes with a square of gummed cloth tape. Using a double strand of the waxed linen thread, make a stitch up from below, over the twig, and back down, tying it securely on the underside. Clip off the extra thread and cover the knot with a piece of gummed paper tape. This makes a nearly invisible, strong reinforcement.

So, there is more than one method of plant mounting, with new methods being tested. Try different ones and find which one is best suited in the facilities available.

For example, for a 4-H project use a notebook and glue the plants into it, or use pages faced with a sheet of clear plastic and mount the plants in that. Loose-leaf pages are best. 4-H instruction manuals suggest taking only part of a plant, not the root, so all the specimens could probably fit into the smaller space.

The important thing is to work carefully, to make the collection as attractive as possible and to have the specimens mounted firmly yet available for study.

Special Problems of Mounting and Packaging

Not all plant specimens are mounted on sheets, and some that are so mounted present special problems.

May apple fruit in a jar

Trees and Other Woody Plants; Fruits

Once the problems of pressing are surmounted, bulky specimens are glued to herbarium sheets in the same manner as herbaceous plants. Dried fruits with a diameter smaller than 1 inch (2.5 cm) can be glued to the sheet, placed in packets, or in little string bags fastened to the sheet. Those fruits larger than an inch in diameter may be removed and stored separately in boxes. One label is glued to the outside of the box, with a duplicate label placed inside. There will usually be a pressed twig or plant to go with each boxed fruit, so make a reference to the fruit on the label of the pressed plant. If there is no pressed portion, a dummy sheet is commonly made—a plain piece of herbarium paper with a label glued in the lower right-hand corner.

Finding cardboard boxes of a convenient size is not always easy, especially for a small herbarium for which large quantities are not required. Most box companies will ship only in lots of 500 or more of each size. You must be ingenious in finding sources, perhaps reusing boxes in which supplies are commonly received, or purchasing some designed for other purposes, such as gift or jewelry boxes, or by buying smaller amounts from a large herbarium. Use boxes just large enough to contain the specimen at hand, not to waste space. Smaller boxes can be grouped inside of a larger one. Boxes can be constructed without great difficulty (see the instructions in Chapter 6).

Fruits may be stored in fluid in jars as described in Chapter 2 with a label inside each jar and a reference either on the pressed specimen or the dummy sheet. Attractive specimen jars may be obtained from biological supply houses, or use simple containers salvaged from other uses. Plastic lids are best because of the corrosive nature of the liquid preservatives. The lids must be absolutely airtight.

Mosses, Liverworts, and Lichens

Dried mosses, liverworts, and lichens may be stored in packets or envelopes. For a packet, fold a sheet of standard-sized typing paper (about 8½ × 11 inches; 21 × 28 cm) into thirds, the folds running parallel to the shorter edge of the paper. Then turn under each side of the folded paper 1½ inches (3.8 cm). Score the fold line and it will turn under more neatly. I use a blunt probe for scoring. Make the score on the side toward which you are folding. The resultant packet will be 5½ inches wide by about 3¾ inches high (14 × 10 cm). Use 100 percent rag-content paper for lasting quality. Instead of making your own, purchase these packets from a biological supply house. A regular herbarium label is pasted to the front of the packet, which is then either fastened to a herbarium sheet or stored card-catalog fashion in a box. Two or more packets of the *same species* may be fastened on a single sheet.

Sometimes mosses, liverworts, or lichens are stored in boxes, or in fluid preservative in jars and vials.

Algae and Other Water Plants

If algae and other water plants were floated out and pressed onto sheets, they will probably need little or no additional gluing, as they are often gelatinous and will glue themselves. If they are on small pieces of paper, each piece may be glued to a standard herbarium sheet, with a label as usual in the lower right-hand corner. If they were floated onto a standard-size paper, simply add the label. Algae that do not glue themselves should be mounted by the usual method. If they are brittle, as the coralline algae are, slip them into a packet and mount the packet on the sheet.

A water plant pressed in the usual way between newspapers (and some of them are perfectly well done this way), may later be found to be very thin and lax so as to be difficult to pick up. In that case bring the paper to the plant. *Planning well first*, paint the back of the plant with glue as it lies on its paper, then lay the herbarium sheet upside down on it. Turn the whole thing over and lift off the newspaper sheet. With careful planning it will be stuck on in the proper place. The nylon-thread grid, described previously in the section, "Spray Gluing, Using a Nylon-Thread Grid," works well for such plants.

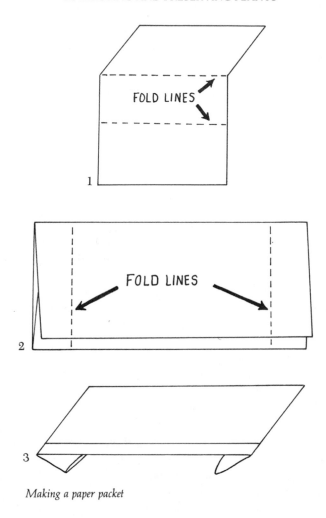

Making a paper packet

Microscopic algae are usually preserved in 50 percent ethyl alcohol or 6-3-1 in small vials, either with plastic screw caps, or in homeopathic or patent-lip vials with rubber or neoprene stoppers, preferably neoprene. Corks may be used, but they tend to permit the fluid to evaporate. If corked vials are used, enclose them in a larger container, with some of the preservative in an open receptacle within this larger container. This will slow the evaporation considerably. Alternatively, dip the caps in melted paraffin wax. Place a small label, with complete data printed very small, inside each vial (see Chapter 4).

Larger algae may be stored in glass jars. Those with straight sides and black plastic caps are attractive. Any set of uniform jars with tight-fitting caps will do, but metal caps may corrode. Again, labels go inside.

Glycerined Algae

Store glycerined algae, used for class demonstration, in tightly closed plastic bags. Handle these carefully, as described in Chapter 2.

Fungi

Fungi, when dried in air or in silica gel, are stored in boxes. One label is glued to the lid or side; a second is placed inside the box. See the section on "Fruits" in Chapter 2.

If the specimen is preserved in fluid, a type of storage jar similar to that used for algae is suitable, with a label inside. FAA is a good general preservative for fungi (see Chapter 2), but for individual species and special purposes, consult the literature.

Microscopic fungi, such as the rusts, are often pressed with the host plant, which is then mounted on a herbarium sheet in the usual way.

Air-Dried Plants for Ornamental Uses

Air-dried plants can be stored in cardboard boxes of suitable size. Purchase florist's boxes or use shoe boxes, or any box of the right size and shape. Be sure to label each one plainly on the outside. Collections of dried plants tend to become unwieldy. The use of boxes of uniform size will help to keep order.

Glycerined Plants for Ornamental Uses

Store glycerined plants in the same way as dried ones.

6

Storage and Display

There is a considerable investment of time and effort in preserving plants. These valuable specimens or ornamental plants deserve careful storage.

Herbarium Sheets

In a herbarium, plants mounted on sheets are stored flat on the shelves of special herbarium cabinets. These cabinets are usually constructed of metal, but may be of wood, and usually contain 26 compartments, each about 19 deep × 13 wide × 8 inches high (48 × 33 × 20 cm). They come in several different styles and can be purchased from biological or herbarium supply houses. They are designed to be airtight so that the fumigants, necessary to prevent insect damage, do not escape.

Species covers

Herbarium sheets bearing plants belonging to a single species are placed within a species cover, a sheet of lightweight paper similar to newsprint, about 16⅝ × 24 inches (42 × 60 cm), folded to 16⅝ × 12 inches (42 × 30 cm). Some herbaria do not use species covers, but their use saves much wear on the plants.

Genus Covers

Species covers are in turn grouped within genus covers, sheets of stiff paper, 17 × 24 inches (43 × 60 cm), folded to 17 × 12 inches (43 × 30 cm). Commonly these genus covers are of jute manila (caliper about 0.015 inch; 0.38 mm). See Chapter 10 for a discussion of the use of different colors for different geographic areas, and for systems of herbarium organization.

Mark each genus cover plainly with the name of the genus contained within it, perhaps stamped in large letters on the outer edge, or printed on a tab fastened to the edge. Write the name of the species on each species cover. I include the family number as well, to aid in refiling specimens.

Family Boards

Within the herbarium cabinets, plants are grouped into families. A family board, a sheet of stiff cardboard the same size as a genus cover, with a tag on the end bearing the family name, may be used to separate one family from another. A card on the front of the cabinet will tell what the groups of plants are inside.

Algae

If algae are mounted on sheets, they are stored in species covers and genus covers.

Collection Books

If mounted sheets are being filed in a notebook, select some reasonable sequence; do not just put them in at random. A small collection could be in alphabetical order by genus and species, or by common name, or even by the time of flowering; however, alphabetical order by family would be preferable. In order to coordinate a collection with a large herbarium, use the sequence given in Chapter 8.

Plants for Ornamental Uses

Plants pressed for ornamental uses may be stored in the papers in which they were pressed. They should be grouped and kept in folders of heavy paper. Be sure to label each folder to make the plants easy to find. Either label them by plant name (asters, delphinium, coral bells); by color (blues, reds, pinks, violets); or by type (vines, tiny plants for stationery, or plants for cartoons). If plants are stored on a shelf, label them with paper strips about 3 × 18 inches (8 × 48 cm). Insert one of these strips in each folder with the caption written on the protruding end.

Pressed plants may also be stored in an old telephone directory, mail-order catalog, or wallpaper-sample book. Fasten index tabs to the edges of the pages to help in finding plants quickly. Organize the plants alphabetically, by color, or by type, as suggested previously.

Plants in Process

There is always a time lag between the collecting and pressing of specimens and their final disposition in the herbarium. During this time, store the plants in the single newspaper sheets in which they were pressed, within folders, in groups of a convenient size. Old genus covers are often used for such folders. The plants of a single collector will probably be stored in order by collection number. Those being set aside for exchange with other herbaria should be stored alphabetically by genus or by family. *Mark every folder* as to its contents. A strip of stiff paper, like that recommended previously for ornamental plants, makes a good marker. Careful recording and marking at every step will save much time lost in searching.

Boxes

Plant specimens in boxes, such as fungi or fruits, should be stored in herbarium cabinets in the same sequence as those on sheets. This poses special problems, since inserting more specimens causes more displacement of the boxes already present. As the collection enlarges, whole groups must be moved along to provide room for the newcomers. It is a good idea to leave an occasional empty compartment to allow for expansion. Another of the difficulties of storing boxes is that they will be of different sizes. A number of smaller boxes can be placed inside a larger one, possibly one the size of the entire shelf.

Fruits

In a small herbarium, simply set aside the top shelves of each cabinet for the storage of fruits that belong to the same families as those housed therein. In a larger herbarium, they will probably be stored in a special section of the room or building. Notations on plant labels, or on the dummy sheets described earlier, will alert users to their presence.

Fungi

Fungi are stored in cabinets according to the filing sequence decided upon (see Chapter 8).

Plants for Ornamental Uses

Plants dried in air or in a dehydrating agent will retain their three-dimensional quality, and take more room to store. Some, like cattails, will have long stems, others such as fungi may be bulky. Florist's boxes make excellent storage containers, and so do shoe boxes. A set of boxes of uniform size will make storage simpler. Set aside some shelves or other special space for a collection. Be sure to label each box plainly and work out an orderly system of arrangement. All dried plants, except the fragrant ones, should be protected by adding a few mothballs or crystals to their storage boxes.

Glycerined plants should be stored in boxes, if they are not to take on a flattened look. Insects will not be much of a problem.

Fragrant Plants

After they are thoroughly dry, store fragrant plants in airtight containers—glass jars, closed plastic bags, and metal or plastic boxes. Do not add moth crystals, which would spoil the fragrance. Many aromatic plants already have an insect-repellant quality.

Packets

Mosses, Liverworts, and Lichens

Storage of packets may be either on herbarium sheets or in boxes. If on sheets, several packets containing the *same species* may be glued to a single sheet, and filed flat in the usual way.

Packets are usually filed in boxes. Shoe boxes are often just the right size to accommodate the packets, and two boxes will fit side by side on the shelf of a herbarium cabinet. Alternatively, boxes may be constructed or purchased in a size to fill the depth of the available shelf.

To construct a box, cut the cardboard, of suitable weight, according to the plan pictured on page 78, score the cardboard along the fold lines on the side toward which you will make the fold, fold it, and glue it at the points of overlap. If the boxes are an uncommon size, it often takes less time to build boxes oneself than to search out a supplier. By making careful plans and measurements, these instructions can be adapted to the making of a box of any size. To make a cover for a box, simply enlarge all the dimensions by the thickness of the cardboard used and use the same plan.

Shipping boxes are often designed exactly to fit the object being shipped. Keep alert for boxes from this source that will suit your

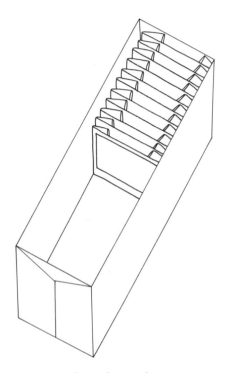

Mosses stored in packets in a box

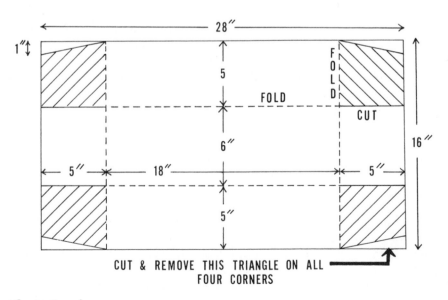

Constructing a box

purposes. If necessary, paint them with quick-drying latex paint for better appearance.

Fumigation

All collections on herbarium sheets or in boxes must be protected from damage by insects such as silverfish or dermestid beetles by being kept constantly under fumigation. Paradichlorobenzene (PDB), moth crystal, is generally used for such routine fumigation. Naphthalene, another kind of moth crystal, is preferred by some workers to kill the variety of insects and larvae that live in fungi. Standard herbarium cabinets have door pockets for the crystals. The PDB or naphthalene must be renewed about three times a year.

For a smaller home collection, place mothballs in a small perforated bag next to the sheets and protect the whole thing within a tightly closed plastic bag.

In a herbarium, an initial fumigation with a stronger chemical is usually carried out before the plants are placed in the cabinets. For a discussion of the methods used, consult the article "Survey of Herbarium Problems," compiled by Thomas B. Croat, in *TAXON*, Journal of the International Association for Plant Taxonomy, Utrecht, Netherlands (May 1978) 27:203–218.

Recent experiments, however, suggest a simple, safe method— placing the plants in a freezer. Deep freezing for about 2 weeks appears to kill most insect pests. If this method fulfills its promise, much handling of hazardous chemicals and their release into the open air by way of exhaust hoods will be avoided.

Vials and Jars

The varying sizes of vials and jars will cause problems, in the same way those of boxes do.

Vials tend to be more uniform in size, since they are used to house microscopic forms. Start by storing them in trays or boxes separated into compartments by dividers. Either purchase the trays and dividers, or make dividers by cutting strips of light cardboard of appropriate size and fitting them together to suit the box at hand (see the diagram on page 80). There are many types of trays on the market. The use of a single tray per genus or species simplifies later rearrangements.

While the materials are in process, which may be a long time, they may be identified by a penciled label *inside* the vial or jar, as well as by the collection number written on a self-adhesive label placed on the stopper

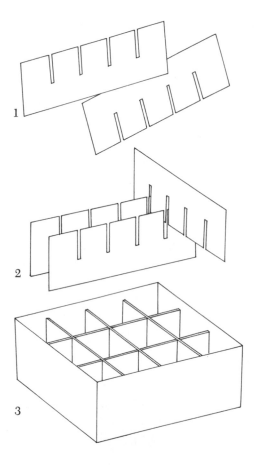

Construction of box dividers

or cap, and stored in order of that collection number. Cleaning the cap
with alcohol helps the labels to adhere better.

Jars may also be stored in trays or boxes, but dividers will only get in
the way as the collection expands, and when jars of different sizes are
inserted among the first ones. Use sizes of trays or boxes that will
multiply to fill the storage shelf efficiently. Four 6 × 9 × 5-inch-high
boxes (15 × 23 × 13 cm) will neatly fill a shelf 19 deep × 13 inches wide
(48 × 33 cm), with room to get your fingers in to extract the boxes.
Alternatively, special large boxes 12 × 8 × 5 inches high (30 × 46 × 13
cm) that would occupy a whole shelf could be bought or constructed.
Individual needs must be studied in order to find the most efficient
system.

Once the specimens are identified and properly labeled as permanent
additions to the collection, store them in the same order as specimens on

herbarium sheets or in boxes. Fruits in jars should be referred to on the label of the associated herbarium specimen or dummy sheet.

Cataloging

A file organized by catalog number calls for a different method, but allows for continuous expansion of a collection without rearrangement.

The catalog of a collection can be on file cards or in a bound book. The bound catalog is harder to lose, but file cards permit easier rearrangement and interfiling if a cross-file is being kept. In both cases the catalog is kept by the *numerical sequence* in which the specimens are added to the collection.

For example, suppose a system for the algae is set up. Numbers could be assigned to each vial of microscopic algae as it is added—1, 2, 3, 4, etc. Or code letters might be assigned to different groups, BG to the blue-green algae, G to the green, etc. The first blue-green alga in the series would be BG-1, the second BG-2, and continuing as specimens are received. In addition to the label within the vial, information on the specimen would be recorded in the catalog under the assigned number. The catalog number would appear on the top or side of the vial and on the permanent label within.

When setting up the catalog on file cards, make carbons or photocopies and have a cross-file by species, locality, and collector. One card, as stated, would have to be filed by catalog number. In order to locate a particular specimen in the collection, consult the catalog file, find its catalog number, and quickly locate the vial.

Scientific Display

At the beginning of this book I pointed out that herbarium specimens are used for research and teaching. The researcher takes the specimens from their folders, works with them, and puts them back carefully. But what if they are used for teaching? How will they survive handling by numbers of students? There are several ways to protect fragile pressed specimens during display, while retaining visibility.

Riker Mounts and Display Boxes

The Riker mount is an ingenious device for displaying specimens. It is essentially a shallow cardboard box filled with cotton and having a glass cover. The specimen, unmounted or mounted, is placed inside the Riker mount and the lid replaced. The plant can then be observed and passed around without damage. For a full-sized herbarium sheet the 12 × 16-inch (30 × 40 cm) Riker mount can be used if the sheet is trimmed a little. When putting together a display specimen, clean the glass. Use a neat label and be sure it is parallel with the case side. Little details like this make the difference between a crude and an attractive demonstration.

Riker mounts come in a series of sizes, from 2½ × 3 × ¾ inches (6 × 8 × 2 cm) to 16 × 24 × ⅞ inches (40 × 60 × 2.2 cm), or even larger and deeper, for displaying various sizes of plants, from a single leaf to a large specimen or set of specimens to demonstrate special characteristics. When preparing something lumpy like pine cones, the deeper kind of Riker mount, or even a glass-topped wooden display case, will be needed.

Botanical Mounts

A botanical mount resembles a Riker mount but has no appreciable depth. It consists of a cardboard back and a glass front, bound together by black tape or passe partout. It comes in about the same dimensions as Riker mounts, and is used to display pressed plants.

Both botanical mounts and Riker mounts are often used to display wall-hung pictures in homes, because they are so attractive.

Transparent Plastic Envelopes

Plastic envelopes 12 × 18 inches in size (30 × 46 cm) may be purchased. The herbarium sheet is inserted in this while being used, then removed and returned to its usual storage.

WHITE SPRUCE.
PICEA CANADENSIS

A Riker mount

All three—Riker mounts, botanical mounts, and plastic envelopes—are available from biological supply houses.

Plastic Laminating

Plastic laminating is a method of preparing sheets permanently for teaching specimens. A special heat press and sheets of Mylar plastic will be needed. The plant, already mounted on herbarium paper with a label, is placed between two sheets of the plastic and inserted in the press for about 1½ minutes, which seals the plastic tightly, making a durable, protected, washable teaching mount.

The press with which I am familiar is made by:

Seal, Inc.
Shelton, Connecticut 06484

Inquiries to that company will bring press sizes, current prices, and local vendors of the presses and the plastic film. This same press can be used for mounting pictures or photographs with photographic mounting tissue, or for laminating reusable captions for displays to be repeated. It is also useful for preparing objects for ornamental uses, explained in Chapter 9.

Let me go into more detail about the steps of laminating. The plants that laminate best are those that are pressed very flat. A stem or twig diameter of ¼-inch (6 mm) will be all right, but more than that will produce a less attractive result.

Preheat the press to a moderate temperature, following the instructions provided by the manufacturer. Cut a piece of the plastic film large enough so that when folded double around the herbarium sheet it will project about ½-inch (13 mm) on the open sides; more than that will wrinkle around the edges; less than that may not quite cover the sheet. Insert the herbarium sheet between the folds of the plastic film, and smooth out all the wrinkles. The static electricity produced will help to hold everything in place. All around the edges, plastic must be touching plastic, as it bonds best to itself.

The plastic-encased sheet then goes into a folder made either of special silicone-treated paper or just plain brown or white paper, larger than the materials that are being mounted. Place the whole assembly in the press, plant side down against the sponge-rubber pad. Always work inside a folder or else the plastic or other matter will get stuck all over the platen of the press, and will then come off on the next thing pressed. Close the press tightly for about 1½ minutes. Experience will teach you the best times and temperatures.

Remove the folder and place it on a flat surface with a large flat weight on it (perhaps a big old book) until it cools—about 1 minute. Trim the edges with sharp scissors (a paper cutter does not work well with plastic), leaving a margin of about $1/16$ inch (2 mm). If you trim away all the margin, the cover may peel.

The result will be a permanent, beautiful, wear-resistant teaching specimen that may be handled over and over by the students, put up for display, stored without preservatives, and washed if it gets soiled.

Quick Plastic Mounts

There are several other methods of preparing display mounts, using various kinds of plastic.

Stationery and bookstores may sell notebook-sized double plastic pages that open up so that the pressed plant specimen may be laid inside. It can be viewed from the front or the back. This is a convenient way for primary or secondary school students, or Scouts or 4-H youths, to display a study collection.

Photographic page mounts with plastic cover sheets may be used for plants. Lay the pressed plant on the base sheet and smooth the plastic over it. Both this mount and the one described previously are only temporary. For longer-lasting displays or collections, the plants should be glued to a backing.

One person made an attractive, uniform set of tree-leaf displays from a set of 9 × 12-inch (23 × 30 cm) brown envelopes. With a razor blade she cut a rectangle from the front of the envelope (inserting a piece of hardboard inside the envelope as a cutting surface), leaving a 1-inch (2.5 cm) frame around the edge. She slipped in a sheet of rigid, clear plastic film, and behind it a card bearing a pressed, labeled specimen of twig and leaves to make a neat mount.

A similar set can be made by mounting labeled tree leaves on a 9 × 12-inch (23 × 30 cm) rectangle of light cardboard, covering it with transparent plastic kitchen wrap or sheet of Mylar plastic, and binding it around the edges with masking tape or colored plastic tape. This makes an inexpensive way to obtain study mounts, not as sturdy as glass or plastic-laminated ones, but adequate for some purposes. Children could use this technique also.

For small displays, try using self-adhesive plastic of the sort used to laminate identification cards, sold in craft and stationery stores. For additional attractiveness, use colored paper backgrounds and frames. Teachers can adapt these ideas for elementary schoolchildren as a combination science, art, and gift project.

Remembering always the amount of time and effort that goes into making a plant collection, whether for science or for ornamental purposes, be sure to give a collection careful storage to prolong its life. By using the methods given for protecting plant specimens during display, some specimens can be brought out of the recesses of the herbarium to make attractive and educational displays, so that students and other viewers are exposed to examples of the real plants, not just pictures in a book. Remember, however, that fading is markedly speeded up in light, so do not expose any scientific specimens to light for any length of time.

7

City Botany

City dwellers wishing to learn to collect and preserve plants and to study botany in the city will be confronted by a vastly different situation from those who live in or close to the open country. Not for the city dweller are the easy field trips to wild forests, bogs, and fields. Although there are the resources of botanical gardens and parks to be enjoyed, collecting there will be sharply restricted or forbidden.

However, even the largest city is not all pavement, but is made up of a series of neighborhoods where people grow trees and shrubs and cultivate flowers and vegetables. Also there are always waste places—vacant lots, railroad yards, waterfronts, canals, and little bits and scraps of uncultivated areas. Within the city plants are to be found that will help the beginner to learn the skills of collecting, preserving, and identifying; and there are plants for ornamental purposes, too. Naturally occurring plants will be useful as scientific specimens.

Weeds are the first allies of the city collector. Tough, persistent, resilient, weeds poke up through every crack in the concrete or blacktop, around porches and building entrances, at the edges of parking lots, in beds of ornamental shrubbery and flowers, and in every waste place. Grasses of many different species can be found. The structure of a grass can be taught or learned perfectly well from a weedy foxtail grass; taken early in development, the heads of that grass are also attractive in dried plant arrangements. Spurges and knotweeds sprawl in tarred parking lots; little blue veronicas, spring-flowering stellarias, and golden dandelions invade lawns. Last summer, along the verge of a Ramada Inn parking lot, in less than 2 hours, I collected 21 different species representing 10 plant families. The plebeian members of a family can represent it as well as the more elegant ones. There is little danger of destroying an endangered species when gathering weeds.

In order to avoid arrest for trespassing, always obtain permission from owners or managers, and scour the waste places of the city. Along railroad tracks and waterfronts you will often find unusual plants brought in by trains and ships, accidentally, from other geographical areas. Any plant that grows wild is of importance as a scientific specimen.

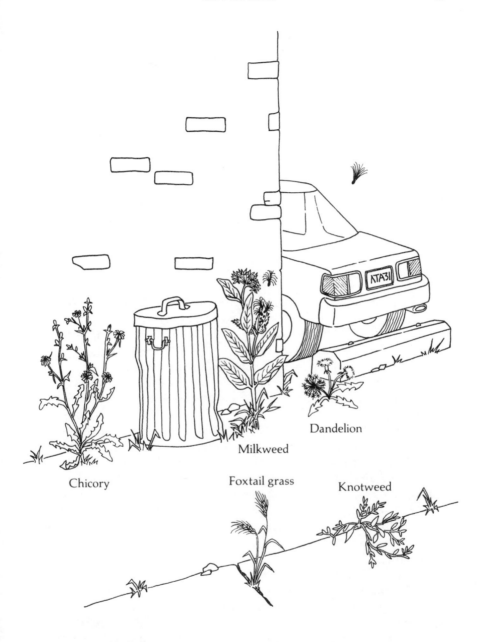

Dandelion

Milkweed

Chicory

Foxtail grass

Knotweed

City lot with weeds
Knotweed: *Polygonum aviculare*
Foxtail grass: *Gramineae*
Chicory: *Cichorium intybus*
Dandelion: *Taraxacum officinale*
Milkweed: *Asclepias* sp.

Butter-and-eggs

A few of the plants to be found growing as weeds in city waste areas are grasses, sedges, and rushes (in damp places), chickweed, campion, carpet weed, euphorbia, knotweed, milkweed, plantain, veronica, dandelion, hawkweed, clover, fleabane, ragweed, thistle, dock, burdock, yarrow, Queen Anne's lace, evening primrose, cinquefoil, butter-and-eggs, wild grape, strawberry, raspberry, and rose. Tree-of-heaven, box-elder, buckthorn, alder, and ash are a few of the common trees and shrubs.

Those who teach botany in its various aspects, and those who demonstrate to groups the ornamental uses of plants, have the responsibility of providing plants for such study and use. Where wild plants are scarce, other sources may be tapped. Some of the suggestions in the following pages will overstep the subject of this book in a stricter sense but will, I hope, be helpful in providing specimens for collection and preservation. These same suggestions can be adapted by both the person collecting for scientific purposes as well as by the one drying plants for ornamental uses.

A second resource in the city will be botanical gardens, parks, and other landscaped areas. Of course, the individual cannot go to these places, digger and pruners in hand, and set to work; but if good relations with the director or overseer are established, friendly cooperation will probably be the result. First of all, they may be glad to part with their weeds. Second, there are often prunings and thinnings available that will be useful as practice subjects for pressing and mounting, or for ornamental purposes.

Closely akin to botanical gardens are greenhouses, conservatories, and florists' shops, possible sources of plants in several ways. Some plants can be purchased outright; prunings and thinnings can be gleaned; there are the weeds that creep in persistently, on and under the benches. *Oxalis* is a regular invader, as are some ferns, liverworts, and easily self-propagated plants. Funeral-home directors are often forced to throw away excess bouquets. Some of these can be used for dissection and study.

Oxalis

Cultivated plants are a source of many materials. Private yards and gardens can supply many specimens, but again, be sure to *ask*!

Just a few of the possibilities among cultivated woody plants are azalea, rhododendron, maple, basswood or linden, lilac, forsythia, flowering crab, dogwood, and evergreens, and from the whole gamut of herbaceous plants that may be cultivated in a particular vicinity; ask for a few or grow your own.

If you are a teacher of botany and the institution where you teach has a greenhouse, you may be able to request that particular plants be grown, or grow them yourself. For scientific study it is important to secure simple flowers. Double ones may fail to show the identifying characteristics of the plant family. For example, the stamens of a flower may have been converted to extra petals by special breeding.

Max Adler, who for some years has struggled with a temperamental greenhouse at Eastern Michigan University at Ypsilanti, gives the following advice (and a glimpse into the problems of greenhouse growing):

> *If* the greenhouse has leaky pipes under the benches, liverworts such as *Lunularia*, *Marchantia*, etc., will thrive, along with "naturalized" *Adiantum* [Maidenhair fern]. Otherwise, use planting trays with field-collected organic soil and clear plastic bag covers. Visit older commercial greenhouses in the area for more weeds.
>
> *If* temperature control is exact along with light duration, then Easter lilies, snapdragons, some roses, chrysanthemums, etc., can be grown. Otherwise purchase them as needed for local greenhouses.

If you have a "misting bench," bedding plants such as alyssum, *Lobelia*, pansy (Johnny-jump-up), *Impatiens*, *Browallia*, *Calceolaria*, etc., will bloom in spring (providing temperature control is on the cool side and a fan provides circulation).

Some plants as *Strelitzia* always bloom in the fall, along with *Solandra*. Others, as *Hymenocallis*, bloom in summer. Corn plant (*Dracaena*) blooms in mid-winter along with poinsettia, Christmas cactus, etc. Figs bear in late spring. *Bouganvillea*, shrimp-plant, Chinese hibiscus, *Rhoeo*, etc., bloom all year. Many large plants must be five years old (*Strelitzia*, *Dracaena*, orchids) to flower. Others need "washtub size" containers to reach blooming size.

"Weeds" include Kenilworth Ivy (spring and summer), *Kalanchoe*, *Oxalis*, ice plant, etc., in ashes surrounding pots. Boston fern, *Tradescantia* and *Zebrina*, *Chlorophytum* (green) are major "escapees" under benches where light intensity is lower. Unfortunately, many greenhouses are designed with hard floors, and pots are kept on elevated slats.

Usually things go a little better if a course is planned by the plants' schedule, not the other way around.

You can study the potentiality and limitations of the greenhouse available to you, and plan accordingly. Given in the following is a list of possible plants to test for greenhouse growing, including those recommended by Max Adler.

Some Greenhouse-Grown Plants to Use in the Teaching of Botany

NOTE: Latin names used in this list are taken from Bailey, *Manual of Cultivated Plants*, 1949, except for those of some orchids, which were supplied by the grower.

Mosses and liverworts and other examples of the more primitive plants can be obtained from biological supply houses, from other greenhouses, or by gathering them from the wild (if they are abundant and not on the protected plant list). They will thrive under benches, in pots, or in damp corners of a greenhouse, or they can be grown on soil rich in humus in planting trays covered with plastic.

Fern fronds

Some Greenhouse-Grown Plants to Use in the Teaching of Botany

General Category	Family	Common Name	Scientific Name	Remarks
Ferns	Polypodiaceae	Boston fern	*Nephrolepis exaltata var. bostoniensis*	Each of the ferns listed shows a different type of sorus or spore-bearing structure. Ferns can be obtained from garden supply houses, biological supply houses, and greenhouses.
		Holly Fern	*Polystichum spp.*	
		Maidenhair fern	*Adiantum pedatum*	
		Polypody fern	*Polypodium vulgare*	
		Staghorn fern	*Platycerium bifurcatum*	
		Table fern; Brake	*Pteris spp.*	
Monocots	Gramineae (Grasses)	Dwarf bamboo	*Sasa veitchii (Sasa pygmaea)*	Most grasses thrive best outdoors, but this one will grow in a greenhouse.
	Cyperaceae (Sedges)	Papyrus	*Cyperus papyrus*	
		Umbrella-plant	*Cyperus alternifolius*	
	Araceae	Dumb cane	*Dieffenbachia seguine*	
		Anthurium	*Anthurium spp.*	
	Bromeliaceae	Spanish moss	*Tillandsia usneoides*	
		Bromeliad	*Billbergia nutans,* etc.	
	Commelinaceae	Moses-in-the-basket	*Rhoeo discolor*	
		Spiderwort	*Tradescantia virginiana*	
		Wandering Jew	*Zebrina pendula*	

General Category	Family	Common Name	Scientific Name	Remarks
	Liliaceae	Corn plant	*Dracaena fragrans*	
		Easter lily	*Lilium longiflorum* var. *eximum*	
	Amaryllidaceae	Spider plant	*Chlorophytum capense*	
		Amaryllis	*Amaryllis vittata*	
		Basket-flower	*Hymenocallis calathina*	
		Spider-lily	*Hymenocallis americana*	
	Iridaceae	Gladiolus	*Gladiolus* spp.	
	Musaceae	Bird-of-paradise	*Strelitzia reginae*	
		Dwarf banana	*Musa nana*	
	Orchidaceae	Orchids	*Cattleya* spp.	
			Cymbidium spp.	
			Dendrobium delicatum	This is small-flowered and easily grown.
			Epidendrum spp.	
			Oncidium ornithorynchum	These grow easily and have many flowers.
			O. ampliatum	
			O. splendidum	
			O. carthaginense	
			Phalaenopsis spp.	
Dicots	Nyctaginaceae	Bougainvillea	*Bougainvillea glabra* var. *sanderiana*	This blooms readily when small.
			Bougainvillea spectabilis	

Family	Common name	Scientific name	Notes
Aizoaceae	Ice plant	Cryophytum crystallinum (Mesembryanthemum crystallinum)	
Fumariaceae (Papaveraceae)	Bleeding heart	Dicentra spectabilis	
Cruciferae (Brassicaceae)	Alyssum	Lobularia maritima	
Crassulaceae	Jade-plant	Crassula argentea	This may not bloom for years.
	Kalanchoe (Bryophyllum)	Kalanchoe daigremontiana	
	Shamrock	Trifolium repens	
Leguminosae (Fabaceae)	Sweet pea	Lathyrus odoratus	
Geraniaceae	Geranium	Perlargonium hortorum	
Oxalidaceae	Oxalis	Oxalis rubra (Oxalis rosea)	This is sometimes called a shamrock, although Trifolium repens is the usual shamrock.
	Sorrel	Oxalis corniculata	A greenhouse weed.
Euphorbiaceae	Chenille plant	Acalypha hispida	
	Croton	Codiaeum variegatum var. pictum	
	Poinsettia	Euphorbia pulcherrima	
Balsaminaceae	Balsam	Impatiens balsamina	
Malvaceae	Chinese hibiscus	Hibiscus rosa-sinensis	
	Flowering maple	Abutilon megapotamicum	

General Category	Family	Common Name	Scientific Name	Remarks
	Violaceae	Johnny-jump-up	*Viola tricolor* var. *hortensis*	
	Begoniaceae	Begonia	*Begonia* spp. especially *B. semperflorens*	
	Cactaceae	Christmas cactus	*Zygocactus truncatus*	
	Ericaceae	Azalea	*Rhododendron* spp.	
	Apocynaceae	Periwinkle	*Vinca rosea*	
	Asclepiadaceae	Ceropegia	*Ceropegia woodii*	
		Wax plant	*Hoya carnosa*	
	Labiatae (Lamiaceae)	Coleus	*Coleus blumei* var. *verschaffeltii*	
	Solanaceae	Browallia	*Browallia* spp.	
		Chalice-vine	*Solandra guttata*	
		Ornamental pepper	*Capsicum frutescens*	
		Tomato	*Lycopersicon esculentum*	
	Scrophulariaceae	Kenilworth ivy	*Cymbalaria muralis*	
		Slipperwort	*Calceolaria integrifolia*	
		Snapdragon	*Antirrhinum majus*	
	Gesneriaceae	African violet	*Saintpaulia ionantha*	
		Cape-primrose	*Streptocarpus kewensis*	
		Gloxinia	*Sinningia speciosa*	
	Acanthaceae	Shrimp-plant	*Beloperone guttata*	
	Campanulaceae	Lobelia	*Lobelia erinus*	
	Compositae	Florist's chrysanthemum	*Chrysanthemum morifolium*	
		Florist's cineraria	*Senecio cruentus* (*Cineraria cruenta*)	

The preceding omits some important families because they do not grow well in a greenhouse. It presents only a few of the species useful for producing fresh plants for winter study.

NOTE: Three excellent articles from the *Michigan Botanist* tell in detail how to grow and force plants, including many of those commonly blooming in the wild in early spring, for use in college-level plant taxonomy classes:

Mellichamp, T. Lawrence. 1976. "Teaching Materials for Botany Classrooms." *Michigan Botanist*, 15:111–122.
Mellichamp, T. Lawrence. 1976. "Forcing Northern Woody Plants into Flower." *Michigan Botanist*, 15:205–214.
Morley, Thomas, and Roberta J. Sladky. 1983. "Winter Taxonomy Class Materials at the University of Minnesota, St. Paul." *Michigan Botanist*, 22:133–140.

Some plants can be "forced" for late-winter bloom. The technique of forcing spring-flowering bulbs is simple. Obtain the bulbs from a garden supply house early in the fall. Plant them in flats or pots so close together they are almost touching and with their tips out of the soil. Bury them, containers and all outdoors or in a cold frame or unheated garage, under about 8 inches (about 20 cm) of sand, sifted ashes, peat moss, sawdust, or straw, and leave them for about 8 to 12 weeks, until the bulbs are well rooted. Bring them into a cool dark place for about a week, then bring them into normal room temperatures. Keep them well watered and they

Amaryllis

will grow and come to flower rapidly. Bulbs useful for forcing are tulips, crocuses, daffodils, narcissus, scillas, grape hyacinths, and amaryllis.

Some perennials that can be forced by similar treatment are *Astilbe*, (spirea), *Hosta* (plantain lily), bleeding heart, and lily of the valley. The bleeding heart will be a city substitute for the white Dutchman's-breeches and squirrel-corn that bloom amid their feathery foliage in the spring woods.

Spring-flowering shrubs also can be forced into bloom after mid-January. Plant the shrubs in pots or tubs in the early fall. Bury the containers to their rims in an ash or sand bed outdoors or in a cold frame. Bring them at first into a cool but not freezing area. After growth begins, bring them into a warm room. Their roots must be kept moist, and their branches should be sprayed several times a day with water. In the home or classroom use a small hand sprayer, such as the kind used for window cleaner. Some of the shrubs that force well are lilac, flowering almond, rhododendron, azalea, mock orange, daphne, heath, Japanese flowering quince, and forsythia.

Instead of using a whole potted shrub, force just the branches of trees and shrubs. Late in the winter, after the plants have endured the period of cold necessary before dormancy can be broken, cut branches up to 3 feet (1 m) long. Bring them into a cool room, about 55–65°F (13–18°C), and plunge the branches immediately into mildly warm water. Mist the branches several times daily with warm water (simulating warm spring rains) and give them good light about 18 hours a day. Pounding the bases of the branches will split apart the fibers and aid in water uptake. In about 4 to 6 weeks, flowers should appear on the branches.

Plants can be grown on window ledges or under fluorescent plant lights in the classroom or in the home. In classes, this method could provide a quadruple lesson: the propagation of plants from seeds or cuttings; the care and culture of plants; the study of flower anatomy as the students dissect some of the flowers produced; and experience in pressing, labeling, and mounting the plant at the end of the school year. Nearly all of these plants can be preserved and used ornamentally. Fifteen reliable plants to grow on a windowsill or under plant lights are listed in the following:

1. Boston fern
2. Wandering Jew
3. Spider plant
4. Small-flowered orchids
5. Kalanchoe
6. Geranium
7. Oxalis
8. Balsam
9. Begonia
10. Periwinkle
11. Coleus
12. Ornamental pepper
13. African violet
14. Chrysanthemum
15. Cineraria

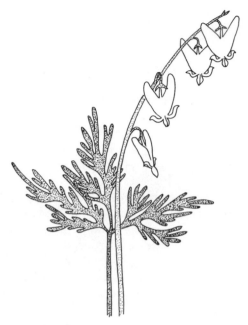

Dutchman's breeches

For the scientific names of these 15 plants, consult the list of greenhouse plants given earlier in this chapter. These are common house plants, and are able to endure a certain amount of neglect.

If a classroom or home is very dry, build a high-humidity area by constructing a wooden frame and covering it on three sides and the top with 4-mil clear plastic. If desired, add a door of similar construction. Shelves built at different heights will provide places nearer to or farther from the light, depending upon the needs of the different plants—less light for ferns; more for balsam.

A terrarium is a useful aid in growing a constant supply of some of the nonflowering plants, such as mosses, liverworts, small ferns, and some fungi. Mix potting soil with charcoal, which will discourage unwanted molds, and pour it over a pebbly layer in the bottom of a glass container. A large fish tank makes a fine terrarium. Obtain the plants from a garden or biological supply house or a greenhouse, or by gathering them from the wild (be sure they are not endangered plants), and set them into the soil. Water them well but do not make the soil too wet. Keep the container covered. In this small enclosed environment, moisture-loving plants will thrive with little additional watering or other care. Give them ample light but not direct sunlight.

If garden space is available, grow plants there to obtain those not readily available elsewhere. Some garden flowers, like petunia and lobelia, can be potted in the fall and kept blooming well into winter. If radishes or broccoli are neglected, they go to flower and will provide examples of crucifers. Weeds, flower and vegetable gardens, shrubs and trees—all can contribute to the teaching or study of botany, and plants to use for ornamental purposes.

By using all the suggested methods, a supply of fresh flowers can be maintained during the winter. A class, garden club, or a 4-H group could begin in the fall with a unit on collecting, identifying, pressing, and mounting tree leaves, or search out and similarly process weedy plants. If the time is not right for keying out the plants, a few flowers may be removed and quick-frozen for later study and the whole plants pressed and mounted. Some cuttings or seeds can be started in the classroom or the home. As winter progresses, the collection and identification of twigs from woody plants will be possible, and perhaps a study in which the bark and the silhouettes of trees and shrubs are photographed. Twigs are usually not pressed for mounting since they tend to shrivel in drying, losing some of their identifying characteristics. Greenhouses, funeral homes, and conservatories may provide flowers for study until spring brings the flowering of early trees and shrubs such as elms, maples, and lilacs, and of the spring-flowering bulbs and early weeds like *Stellaria*. Plants brought to flower by forcing will also provide fresh material. The quick-freezing method mentioned in Chapter 2 may be used to insure a supply of flowers from other families.

Spore-bearing plants—mosses, liverworts, ferns, and fungi—can be cultured in terrariums. Watching and waiting for fruiting bodies to develop on these plants will help to fix in your mind the plants' life cycles.

Exposure to a multitude of plant families is not necessary for a beginning botanist. The characteristics of plant families and the correct techniques of collecting, identifying, pressing, and mounting can be learned by using whatever is available. Once the correct techniques are learned, they can be applied in higher studies.

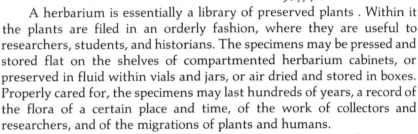

8

The Herbarium
—An Introduction

A herbarium is essentially a library of preserved plants . Within it the plants are filed in an orderly fashion, where they are useful to researchers, students, and historians. The specimens may be pressed and stored flat on the shelves of compartmented herbarium cabinets, or preserved in fluid within vials and jars, or air dried and stored in boxes. Properly cared for, the specimens may last hundreds of years, a record of the flora of a certain place and time, of the work of collectors and researchers, and of the migrations of plants and humans.

Herbaria are organized for different purposes. Very small ones may represent the plants present in a single locality, like a botanical garden or a research woodlot. Some contain all the collections of one person, or voucher specimens attesting to the work done on a particular study. Larger ones attempt to cover the flora of a whole state or country, or of the world. The specimens may serve as teaching aids or as reference and research collections. For greatest usefulness to botanical science, specimens should be stored in herbaria with facilities for processing loans, where they will be accessible to researchers.

Each herbarium is arranged according to a system. Some are simply arranged alphabetically by family, then alphabetically by genus and species within that family. Most follow a phylogenetic sequence, arranging the families in order from the more primitive to the more advanced, according to the best-informed opinions at the time the system was devised. There is more than one such system. The sequence offered at the end of this chapter is a supplemented version of the worldwide system of Engler and Prantl, used in many herbaria.

In large collections, those plants of the immediate geographic area are often placed in separate folders at the head of each genus grouping, and genus covers of different colors are used to denote the major geographical areas from which the plants come. For example, specimens of *Rosa* from Michigan might be housed in green folders placed first, those of the rest of the United States and Canada in buff-colored folders, with blue for Latin American, yellow for Asia, orange for Africa, gold for Europe, and turquoise for Australia and the Pacific islands, as was done in one herbarium. Only a large collection justifies such separation, but it does make easier the task of a researcher looking for plants of a certain part of the world.

Difficult specimens identified only to family may be placed at the beginning or end of each family; those identified only by genus at the beginning or end of each genus. Some expert may later consult the collection and identify those plants. Most herbaria do not guarantee accuracy. Plants are filed under the name assigned them as they came in. As experts consult the collections, they make determinations and corrections. However, workers strive for a high degree of accuracy.

The time and labor required to produce a good herbarium specimen

would make the building of collections prohibitively expensive in most cases if herbaria had to buy their specimens. Collections are usually built up of plants collected by university staff researchers and by students, or by gifts and exchanges between herbaria. Sometimes experts will identify plants for others in exchange for duplicate specimens.

When sending plants for identification, exchange, or gifts, send them unmounted, just in their pressing papers, with a good label on each. Every herbarium has its own mounting methods and choice of papers, and would prefer to do the mounting. It is a courtesy to send labels made on 100 percent rag content paper. Photocopies on this paper are acceptable, although originals are better. Never send carbon copies.

The equipment needed in a herbarium may be purchased from biological supply houses or special herbarium equipment sources. Should you find yourself in charge of a small herbarium, or assigned to build up a

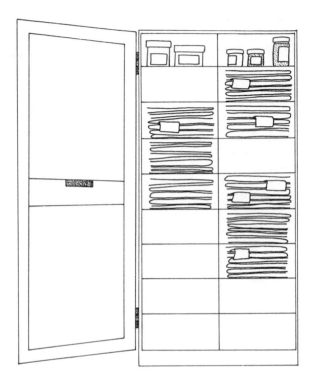

Herbarium cabinet with plants in folders and boxes

collection, consult the curators of other herbaria for the current addresses of such companies.

The details of herbarium storage were discussed in Chapter 6.

Accessioning and Cataloging

Curators of herbaria keep a running record of additions to the collections. As each group of plants comes in, the curator assigns each plant an accession number, and records the accession. The entry in the accession records might read something like this: "August 1, 1984. 150 flowering plants in papers. Received as an exchange from Ruth B. MacFarlane, Michigan Technological University, Ford Forestry Center Herbarium, L'Anse, Michigan. Plants of Ontonagon County. Accession numbers 60,001 through 60,150."

In some herbaria, a single accession number is assigned to the whole group of plants received at one time, so that all the above plants might be recorded as accession number 60,001.

The accession numbers may be written on the plant labels immediately, or the whole set of plants may be stamped with accession numbers when they have been mounted and are ready for storage.

In many herbaria, a card catalog is kept of all plant specimens. File cards are made out, containing all the label data, and kept as a reference file. They may be filed in order by accession number, by species, by collector, by locality, or all of these. Such a record makes it easier to compile data on the collections without undue wear on the plants. A number of large herbaria have abandoned the practice, because of the clerical time necessary. However, many herbaria are computerizing the data from their collections, which makes retrieval of the stored information possible in a variety of ways useful to the researchers. A not impossible goal is to have all the herbaria of the world recorded by computer.

More details of herbarium management are beyond the scope of this book. The foregoing discussion was intended just as an introduction to the herbarium. But I should like to make a few suggestions on the use, and the etiquette, of the herbarium.

Before seeking to use the resources of the herbarium, be competent in the use of the microscope, and be at least familiar with the techniques of research.

As a beginning student–researcher, approach the herbarium with respect. Here lie countless hours of work and study, and irreplaceable specimens. There is no going back even 10 years in time to collect at a site, and many sites have been changed irrevocably under the plow, the chainsaw, and the bulldozer. Where once queen-of-the-prairie grew, asphalt now covers a parking lot, and drainage ditches have emptied swamps that were formerly the home of the wild orchid.

Ask the curator before delving into the collections. Ask if there are any prohibitions or restrictions. Ask for advice on handling the specimens, even if you are sure you know what you are doing. Use only a pencil for taking notes, to avoid making an indelible mark on the herbarium sheet. Ask before refiling plants; the curator may prefer to file them him- or herself. A misfiled plant is a lost one.

Anyone studying a particular group of plants will be expected to report his or her findings to the herbarium. For example, if a researcher corrects the identification given on the label, instead of scrawling his or her determination on the sheet, a narrow annotation label should be affixed to the sheet, just above the present label if possible. The researcher's full name and the year should be included. You may be so well known in botanical circles that you are recognized by your initials today, but a hundred years from now you may be a puzzle.

Smoking, food, and beverages have no place in the herbarium, with their potential for damaging the sheets. If a researcher is working long hours, and needs coffee, the cup should be placed on some surface other than the worktable, so that the contents cannot possibly be spilled onto the specimens.

Pressed plant specimens are brittle, and not all are mounted firmly. Rough handling will break off bits and pieces. If a piece does break off, it should be saved by placing it in a fragment folder attached to the sheet. If there is no such folder, the researcher is responsible for obtaining (or improvising) one. Clip it to the herbarium sheet until the curator can glue it on properly.

Keep the specimens right side up. *Never* turn them over like pages in a book. Be sure the sequence of filing is understood and keep the specimens in the same order. Replace specimens in their folders neatly, so that no edge or corner of a sheet projects.

If a flower must be removed from a herbarium specimen file for study, the researcher must be sure that his or her competence is of a high enough order to handle it. After dissecting and studying it, preserve it carefully in a folded square of waxed paper and place it in the fragment folder. Dissect the specimen on a glass slide. After completing the dissection, cover the specimen with a few drops of warm glycerin and store the slide in a flat, dustproof case, labeled to identify it with the herbarium specimen from which it was taken. Occasionally, during the storage, add a drop of a 1 percent solution of oxyquinoline sulfate. Some of the newer, quick-mounting methods would substitute for this method.

Respect the research time of others. Do not listen to a radio while working unless you are alone or when wearing earphones. Do not chat with other researchers. This may be a crucial study period for the other person, with time running out on a deadline.

Finally, many classes in plant systematics no longer require student collections, because of class size or because they are taught at city universities. A student who becomes interested in doing continuing work in plant taxonomy, studying a particular group of plants, or even revising a genus, will need to study great numbers of plants, some of them collected by others, some by him- or herself. A student will need to know how to properly collect and preserve specimens. Not every professor is well trained in these techniques, so that a student may need to learn the methods for him- or herself.

For convenience in describing plant relationships, plants are divided into smaller and smaller categories until the individual is reached. Beginning with the whole plant kingdom, plants are classified according to Division, followed by Class, Order, and so on down. These relationships are shown in the following chart.

Each taxonomic rank (or taxon) contains all the ones below it, i.e., each genus can be in only one family, but each family may comprise numerous genera.

Taxonomic Hierarchy of Plants

Kingdom
  Division
    Class
      Order
        Family
          Species

There can be graduations of these, as Subclass, Suborder, Superfamily, Subfamily, Subspecies, Variety.

Sample Hierarchy

Kingdom: Plantae (all plants)
  Division: Magnoliophyta (Flowering plants or Angiosperms)
    Class: Liliopsida (Monocotyledons)
      Order: Liliales (Lily order)
        Family: Liliaceae (Lily family)
          Genus: *Lilium* (Lily genus)
            Specific epithet: *michiganense* Farw.

Thus, the scientific name of the Michigan lily is *Lilium michiganense* Farw.

Herbarium Filing Systems

Within the herbarium, specimens are filed according to some orderly sequence. Here are some suggestions for filing the different plant groups.

Algae

The Algae may be filed by division as follows:

- Cyanophyta (Blue-green algae)
- Chlorophyta (Green algae)
- Charophyta (*Chara*, etc.)
- Phaeophyta (Brown algae)
- Chrysophyta (Golden algae, including the Diatoms and *Vaucheria*)
- Euglenophyta (Euglenoids)
- Cryptophyta (Cryptomonads)

Within each division specimens may be filed alphabetically by genus.

Fungi

The Fungi may be filed according to the system given in Ainsworth and Bisby's *Dictionary of the Fungi*, Sixth Edition (Commonwealth Mycological Institute, Kew, Surrey, England, 1971) and Rolf Singer's *The Agaricales in Modern Taxonomy* (J. Cramer, 1975). They are filed by class within each division, by order within each class, alphabetically by family within the order, then alphabetically by genus and by species within the family.

Liverworts, Mosses, and Lichens

Liverworts (Hepatophyta), Mosses (Bryophyta), and Lichens may be filed alphabetically by genus within those three broad classifications. Lichens, being a symbiotic association of algae and fungi, are often grouped with the Fungi.

Ferns and Fern Allies (Vascular Cryptogams)

The filing sequence given in the following for the ferns is a conservative one, and the family members are purely arbitrary choices. Research in plant taxonomy continues, and rearrangements are continually being made.

Flowering Plants (Spermatophyta)

The Herbarium Filing Sequence for the flowering plants is based on the Englerian system, used in many of the herbaria of the world, with additions from Lawrence (*Taxonomy of Vascular Plants*, 1951), Willis (*A Dictionary of the Flowering Plants and Ferns*. 6th ed., 1966), and from occasional practical decisions. Following each family name in the Filing Sequence is either the common or the scientific name of a representative genus.

Herbarium Filing Sequence (Ferns and Their Allies)

A-1 Equisetaceae
A-2 Lycopodiaceae
A-3 Selaginellaceae
A-4 Psilotaceae
A-5 Isoetaceae
A-6 Ophioglossaceae
A-7 Marattiaceae
A-8 Osmundaceae
A-9 Schizaeaceae

A-10 Gleicheniaceae
A-11 Hymenophyllaceae
A-12 Cyatheaceae
A-13 Dicksoniaceae
A-14 Polypodiaceae
A-15 Parkeriaceae
A-16 Marsileaceae
A-17 Salviniaceae

Herbarium Filing Sequence
(Seed Plants or Flowering Plants)

Gymnosperms

1. Cycadaceae. Cycad
2. Bennettiaceae. Bennettia (extinct)
3. Cordaitaceae. Cordaites (extinct)
4. Ginkgoaceae. Ginkgo
5. Taxaceae. Yew
5a. Podocarpaceae. Podocarpus
5b. Araucariaceae. Araucaria
5c. Cephalotaxaceae. Cephalotaxus
6. Pinaceae. Pine
6a. Taxodiaceae. Taxodium
6b. Cupressaceae. Cypress
7. Ephedraceae. Ephedra
7a. Gnetaceae. Gnetum
7b. Welwitschiaceae. Welwitschia

Angiosperms: Monocots

8. Typhaceae. Cattail
9. Pandanaceae. Screw pine
10. Sparganiaceae. Bur reed
11. Potamogetonaceae (Zosteraceae). Pond weed
11a. Cymodoceaceae. Cymodocea
12. Najadaceae. Najas
13. Aponogetonaceae. Aponogeton
14. Juncaginaceae (Scheuchzeriaceae). Arrow grass
15. Alismataceae (Alismaceae). Water plantain
16. Butomaceae. Flowering rush
17. Hydrocharitaceae. Frog's bit
18. Triuridaceae. Triurus
19. Gramineae (Poaceae). Grass
20. Cyperaceae. Sedge
21. Palmae (Arecaceae). Palm
22. Cyclanthaceae. Panama-hat palm
23. Araceae. Arum
24. Lemnaceae. Duckweed
25. Flagellariaceae. Flagellaria
26. Restionaceae. Restio

27. Centrolepidaceae.   Centrolepis
28. Mayacaceae.   Bog moss
29. Xyridaceae.   Yellow-eyed grass
30. Eriocaulaceae.   Pipewort
30a. Thurniaceae.   Thurnia
31. Rapateaceae.   Rapatea
32. Bromeliaceae.   Pineapple
33. Commelinaceae.   Spiderwort
34. Pontederiaceae.   Pickerel weed
34a. Cyanastraceae.   Cyanastrum
35. Philydraceae.   Philydrum
36. Juncaceae.   Rush
37. Stemonaceae.   Stemona
38. Liliaceae.   Lily
39. Haemodoraceae.   Bloodwort
40. Amaryllidaceae.   Amaryllis
41. Velloziaceae.   Vellozia
42. Taccaceae.   Tacca
43. Dioscoreaceae.   Yam
44. Iridaceae.   Iris
45. Musaceae.   Banana
46. Zingiberaceae.   Ginger
47. Cannaceae.   Canna
48. Marantaceae.   Arrowroot
49. Burmanniaceae.   Burmannia
50. Orchidaceae.   Orchid

Dicots

51. Casuarinaceae.   Casuarina
52. Saururaceae.   Lizard's tail
53. Piperaceae.   Pepper
54. Chloranthaceae.   Chloranthus
55. Lacistemaceae.   Lacistema
56. Salicaceae.   Willow
57. Myricaceae.   Bayberry
58. Balanopsidaceae.   Balanopsis
59. Leitneriaceae.   Cork wood
60. Juglandaceae.   Walnut
60a. Julianiaceae.   Juliania
61. Betulaceae.   Birch
62. Fagaceae.   Beech
63. Ulmaceae.   Elm
63a. Rhoipteleaceae. Rhoiptelea

64. Moraceae. Mulberry
65. Urticaceae. Nettle
66. Proteaceae. Protea
67. Loranthaceae. Mistletoe
68. Myzodendraceae. Myzodendron
69. Santalaceae. Sandalwood
70. Grubbiaceae. Grubbia
71. Opiliaceae. Opilia
71a. Ocktoknemataceae. Octoknema
72. Olacaceae. Olax
73. Balanophoraceae. Balanophora
74. Aristolochiaceae. Birthwort
75. Rafflesiaceae. Rafflesia
76. Hydnoraceae. Hydnora
77. Polygonaceae. Buckwheat
78. Chenopodiaceae. Goosefoot
78a. Didiereaceae. Didierea
79. Amaranthaceae. Amaranth
80. Nyctaginaceae. Four-o'clock
81. Batidaceae. Batis
82. Theligonaceae. (Cynocrambaceae) Theligonum
83. Phytolaccaceae. Pokeweed
83a. Gyrostemonaceae. Gyrostemon
83b. Achatocarpaceae. Achatocarpus
84. Aizoaceae. Mesembryanthemum
85. Portulacaceae. Purslane
86. Basellaceae. Basella
86a. Dysphaniaceae. Dysphania
87. Caryophyllaceae. Pink
88. Nymphaeaceae. Water lily
89. Ceratophyllaceae. Hornwort
90. Trochodendraceae. Trochodendron
90a. Cercidiphyllaceae. Cercidiphyllum
91. Ranunculaceae. Buttercup
92. Lardizabalaceae. Lardizabala
92a. Sargentodoxaceae. Sargentodoxa
93. Berberidaceae. Barberry
93a. Leonticaceae. Leontice
94. Menispermaceae. Moonseed
95. Magnoliaceae. Magnolia
95a. Himantandraceae. Himantandra
95b. Tetracentraceae. Tetracentron
95c. Degeneriaceae. Degeneria
96. Calycanthaceae. Calycanthus

97. Lactoridaceae. Lactoris
98. Annonaceae. Custard apple
98a. Eupomatiaceae. Eupomatia
99. Myristicaceae. Nutmeg
100. Gomortegaceae. Gomortega
101. Monimiaceae. Monimia
102. Lauraceae. Laurel
103. Hernandiaceae. Hernandia
104. Papaveraceae. Poppy
104a. Fumariaceae. Fumitory
105. Cruciferae. Mustard
106. Tovariaceae. Tovaria
107. Capparidaceae. Caper
108. Resedaceae. Mignonette
109. Moringaceae. Moringa
109a. Bretschneideraceae. Bretschneidera
110. Sarraceniaceae. Pitcher plant
111. Nepenthaceae. Nepenthes
112. Droseraceae. Sundew
113. Podostemaceae. Riverweed
113a. Tristichaceae. Tristicha
114. Hydrostachydaceae. Hydrostachys
115. Crassulaceae. Orpine
116. Cephalotaceae. Cephalotus
117. Saxifragaceae. Saxifrage
117a. Grossulariaceae. Gooseberry
118. Pittosporaceae. Pittsporum
118a. Roridulaceae. (Byblidaceae. Byblis) Roridula
119. Brunelliaceae. Brunellia
120. Cunoniaceae. Cunonia
121. Myrothamnaceae. Myrothamnus
122. Bruniaceae. Brunia
123. Hamamelidaceae. Witch hazel
123a. Roridulaceae. Roridula
123b. Eucommiaceae. Eucommia
124. Platanaceae. Plane tree
125. Crossosomataceae. Crossosoma
126. Rosaceae. Rose
126a. Chrysobalanaceae. Chrysobalanus
127. Connaraceae. Connarus
128. Leguminosae (Fabaceae). Pea
128a. Pandaceae. Panda
129. Geraniaceae. Geranium
130. Oxalidaceae. Wood sorrell

131. Tropaeolaceae.   Tropaeolum
132. Linaceae.   Flax
133. Humiriaceae.   Humiria
134. Erythroxylaceae.   Coca
135. Zygophyllaceae.   Caltrop
136. Cneoraceae.   Cneorum
137. Rutaceae.   (Rhabdaceae. Rhabdodendron) Rue
138. Simaroubaceae.   Quassia
138a. Surianaceae.   Suriana
139. Burseraceae.   Bursera
140. Meliaceae.   Melia
140a. Akaniaceae.   Akania
141. Malpighiaceae.   Malpighia
142. Trigoniaceae.   Trigonia
143. Vochysiaceae.   Vochysia
144. Tremandraceae.   Tremandra
145. Polygalaceae.   Milkwort
145a. Krameriaceae.   Krameria
146. Dichapetalaceae.   Dichapetalum
147. Euphorbiaceae.   Spurge
147a. Daphniphyllaceae.   Daphniphyllum
148. Callitrichaceae.   Water starwort
149. Buxaceae.   Buxus
150. Empetraceae.   Empetrum
151. Coriariaceae.   Coriaria
152. Limnanthaceae.   False mermaid
153. Anacardiaceae.   Cashew
154. Cyrillaceae.   Cyrilla
155. Pentaphylacaceae.   Pentaphylax
156. Corynocarpaceae.   Corynocarpus
157. Aquifoliaceae.   Holly
158. Celastraceae.   Staff tree
158a. Dipentodonaceae.   Dipentodon
159. Hippocrateaceae.   Hippocratea
160. Stackhousiaceae.   Stackhousia
161. Staphyleaceae.   Bladdernut
162. Icacinaceae.   Icacina
162a. Aextoxicaceae.   Aextoxicon
163. Aceraceae.   Maple
164. Hippocastanaceae.   Horse chestnut
165. Sapindaceae.   Soapberry
166. Sabiaceae.   Sabia
167. Melianthaceae.   Melianthus
168. Balsaminaceae.   Touch-me-not

169. Rhamnaceae. Buckthorn
170. Vitaceae. Grape
171. Elaeocarpaceae. Elaeocarpus
172. Sarcolaenaceae. (Chlaenaceae) Sarcolaena
173.
174. Tiliaceae. Linden
175. Malvaceae. Mallow
176.
177. Bombacaceae. Bombax
178. Sterculiaceae. Sterculia
179. Scytopetalaceae. Scytopetalum
180. Dilleniaceae. Dillenia
180a. Actinidiaceae. Actinidia
181. Eucryphiaceae. Eucryphia
181a. Medusagynaceae. Medusagyne
182. Ochnaceae. Ochna
182a. Strasburgeriaceae. Strasburgeria
183. Caryocaraceae. Caryocar
184. Marcgraviaceae. Marcgravia
185. Quiinaceae. Quiina
186. Theaceae. Tea
187. Hypericaceae (Guttiferae). St. Johnswort
188. Dipterocarpaceae. Dipterocarpus
189. Elatinaceae. Waterwort
190. Frankeniaceae. Frankenia
191. Tamaricaceae. Tamarisk
192. Fouquieriaceae. Candlewood
193. Cistaceae. Rock-rose
194. Bixaceae. Bixa
195. Cochlospermaceae. Cochlospermum
196. Koeberliniaceae. Koeberlinia
197. Canellaceae (Winteranaceae). Wild cinnamon
198. Violaceae. Violet
199. Flacourtiaceae. Flacourtia
200. Stachyuraceae. Stachyurus
201. Turneraceae. Turnera
202. Malesherbiaceae. Malersherbia
203. Passifloraceae. Passion flower
204. Achariaceae. Acharia
205. Caricaceae. Carica
206. Loasaceae. Loasa
207. Datiscaceae. Datisca
208. Begoniaceae. Begonia
209. Ancistrocladaceae. Ancistrocladus

210. Cactaceae.   Cactus
211. Geissolomataceae.   Geissoloma
212. Penaeaceae.   Penaea
213. Oliniaceae.   Olinia
214. Thymelaeaceae.   Mezerum
215. Elaeagnaceae.   Oleaster
216. Lythraceae.   Loosestrife
216a. Heteropyxidaceae.   Heteropyxis
217. Sonneratiaceae.   Sonneratia
217a. Crypteroniaceae.   Crypteronia
218. Punicaceae.   Pomegranate
219. Lecythidaceae.   Lecythis
219a. Barringtoniaceae.   Barringtonia
220. Rhizophoraceae.   Mangrove
220a. Alangiaceae.   Alangium
221. Combretaceae.   Combretum
222. Myrtaceae.   Myrtle
223. Melastomaceae.   Melastoma
223a. Trapaceae.   (Hydrocaryaceae) Water-chestnut
224. Onagraceae.   Evening primrose
225. Haloragaceae.   Water milfoil
225a. Hippuridaceae.   Hippuris
226. Cynomoriaceae.   Cynomorium
227. Araliaceae.   Ginseng
228. Umbelliferae.   (Apiaceae) Parsley
229. Cornaceae.   Dogwood
229a. Garryaceae.   Garrya
229b. Nyssaceae.   Nyssa
230. Clethraceae.   Pepperbush
231. Pyrolaceae.   Pyrola
232. Lennoaceae.   Lennoa
233. Ericaceae.   Heath
234. Epacridaceae.   Epacris
234a. Theophrastaceae.   Joewood
235. Diapensiaceae.   Diapensia
236. Myrsinaceae.   Myrsine
237. Primulaceae.   Primrose
238. Plumbaginaceae.   Leadwort
239. Sapotaceae.   Sapodilla
239a. Sarcospermataceae.   Sarcosperma
239b. Hoplestigmataceae.   Hoplestigma
240. Ebenaceae.   Ebony
240a. Oncothecaceae.   Oncotheca
240b. Diclidantheraceae.   Diclidanthera

241. Styracaceae.   Styrax
241a. Lissocarpaceae.   Lissocarpus
242. Symplocaceae.   Symplocos
243. Oleaceae.   Olive
243a. Desfontaineaceae.   Desfontainea
244. Salvadoraceae.   Salvadora
245. Loganiaceae.   Logania
246. Gentianaceae.   Gentian
247. Apocynaceae.   Dogbane
248. Asclepiadaceae.   Milkweed
249. Convolvulaceae.   Morning glory
250. Polemoniaceae.   Phlox
250a. Cobaeaceae.   Cobaea
251. Hydrophyllaceae.   Waterleaf
252. Boraginaceae.   Borage
252a. Ehretiaceae.   Ehretia
253. Verbenaceae.   Vervain
253a. Symphoremataceae.   Symphorema
254. Labiatae (Lamiaceae).   Mint
254a. Tetrachondraceae.   Tetrachondra
255. Nolanaceae.   Nolana
256. Solanaceae.   Nightshade
257. Scrophulariaceae.   Figwort
258. Bignoniaceae.   Trumpet creeper
259. Pedaliaceae.   Pedalium
260. Martyniaceae.   Unicorn plant
261. Orobanchaceae.   Broomrape
262. Gesneriaceae.   Gesneria
263. Columelliaceae.   Columellia
264. Lentibulariaceae.   Bladderwort
265. Globulariaceae.   Globularia
266. Acanthaceae.   Acanthus
266a. Mendonciaceae.   Mendoncia
267. Myoporaceae.   Myoporum
268. Phrymaceae.   Lopseed
269. Plantaginaceae.   Plantain
270. Rubiaceae.   Madder
271. Caprifoliaceae.   Honeysuckle
272. Adoxaceae.   Adoxa
273. Valerianaceae.   Valerian
274. Dipsacaceae.   Teasel
275. Cucurbitaceae.   Gourd
276. Campanulaceae.   Bellflower
276a. Lobeliaceae.   Lobelia

277. Goodeniaceae.   Goodenia
277a. Brunoniaceae.   Brunonia
277b. Stylidiaceae.   Stylidium
278. Candolleaceae.   Candollea
279. Calyceraceae.   Calycera
280. Compositae.   (Asteraceae). Daisy   Aster

9

Ornamental Uses —Two-Dimensional

Exploring the ornamental uses of dried plants opens the door to a new world of artistic adventure where the beauty of color and shape in every plant can be seen, even in common weeds. As more is learned about design and technique, the individual's vision continues to expand endlessly. (Appendix 2 lists plants used ornamentally.)

We shall explore the ornamental uses of plants under four main topics: two-dimensional pictures and two-dimensional decorations in this chapter; three-dimensional arrangements and scented plants in Chapter 10.

Elements of Two-Dimensional Design

Before beginning to work with pressed and dried plants, stop a moment to consider the elements of design. There should be a balance of colors, textures, light and dark areas, solids and spaces, mass and line. Balance, harmony, scale, repetition, focal point, rhythm, and unity must all be considered. There are no rigid rules, but flower arranging is an art, and an awareness of the elements of design will produce more pleasing results.

At first, this discussion may not seem clear, but read through it once, then keep coming back to it while working with the actual plants; it will help in making decisions about the colors and shapes at hand.

Balance insures that a design is not lopsided or top heavy. Dark colors appear heavier than light colors, so they should be placed nearer the center or the base of the grouping. Smooth leaves and petals reflect light, thus appearing larger. Rough surfaces absorb light, appearing smaller. When balancing size, remember that a smaller element placed farther from the center will balance a larger one placed nearer the center. Grouped small flowers will balance a single large one. An arrangement may be symmetrical or asymmetrical, but imagine a line through the

115

center of the composition. Is the weight properly distributed? For example, for those three pink flowers to the left of the center line, is there a maroon one to the right of it?

To achieve *harmony*, make sure that colors and shapes combine well with one another and do not clash. A brilliant bird-of-paradise flower would not be at home in a cloud of baby's breath, but would combine happily with the bold shapes of canna leaves.

Scale is the relationship of height to breadth, and of the size of the components to the size of the arrangement and to the background and edging materials. Delicate, pale plants need a finer grained background and would be overpowered by a massive frame, which might be just right for a bold arrangement of brilliant flowers.

Repetition means the use of the same color or shape again and again.

A *focal point* should be created, a place where the eye comes to rest. The lines of the rest of the design will appear to radiate from it. The focal point can be the brightest or largest flower or leaf, or the largest area of small or light-colored flowers, or an unusual shape.

Rhythm might be achieved by repeating units in threes: three light-blue flowers, three dark-blue flowers, six yellow ones, for example. If a vine has been pressed, make use of its S-curves and circles to create rhythm in the design. Let it loop and curve, laying out its leaves to left and right, letting tendrils make curlicues of decoration that repeat the curves of the vine.

Does the arrangement have *unity*? Does one color or shape dominate or neatly balance another? Does the composition have an overall form? The elements can be filled into an imaginary triangle, oval, crescent or C, open circle, heart shape, a fan, an L, an elegant S-curve, or elongated vertical and horizontal arrangements.

Spaces must be left so that every component can be seen clearly. It adds interest if little sparks of surprises are introduced—a change of shape or color—but these should not disrupt the unity of the whole. Designs with graceful rhythm or free-flowing lines will make the best use of the naturalness of plants. Sometimes all lines will flow to one side of the picture, with a large flower as a center of balance, or they may branch from the central axis like the limbs of a tree.

Background materials should enhance the shapes, textures, and colors of the plants used, and not dominate them. Remember that the flowers are of primary interest. However, the background can be the important element in contrast, adding richness and drama. Materials to be used include watercolor paper, blotting paper, rice paper, silk, velvet, linen, and burlap. The best colors are black, white, and cream; avoid beige, for if the plant colors fade in time, they will tend to become pale tan and blend with the background. Even faded plants remain attractive, however, because of their shapes. Silver and white plants look attractive against

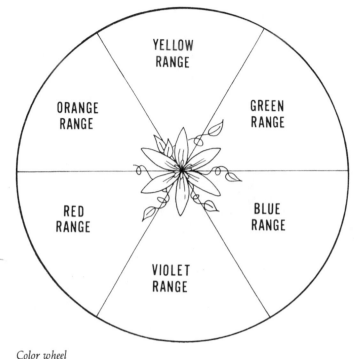

Color wheel

red or blue; golden grasses and yellow, orange, and white flowers show up on black.

Frames should fit the mood of the arrangement. They may be oval, round, rectangular, and elongated–rectangular. Plain ones help to bring out the beauty of the plants; ornate ones detract. Use a silver frame with a black velvet background, a white frame with pastel colors, or modern plain wood with most plants.

Some *themes* might be a set of four pictures in oval frames, typifying spring, summer, fall, and winter; or a pair of pictures representing the flowers of forest and field; or a single composition showing the life cycle of a single plant from seedling through flowers, leaves, and fruit, perhaps containing a skeletonized leaf.

Elements of Color

The use of color is a vital part of designing with flowers. The understanding of some terms used to describe color will be helpful. *Hue* is the name of a color, such as red, yellow, and blue. *Value* is its lightness or darkness, *chroma* its intensity. A *tint* is a color made lighter by the addition of white, a *shade* is a color made darker by the addition of black, and a *tone*

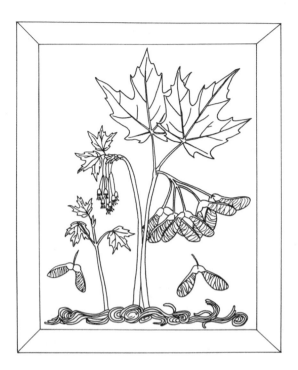

Framed life cycle of maple

is a color made duller by the addition of gray.

Warm colors, red and yellow, are advancing. Reds are vital and warm, yellows are sunny and cheerful, and orange is gay and warm. *Cool colors*, blues and greens, are receding. Purple has spiritual appeal, pink and blue are dainty and gentle, and brown is earthy, dependable, and steady.

The combinations in which colors are used are important. In an open field, nature can mix colors with an indiscriminate hand, but in the confined area of a design or picture, selectivity is needed (see the color wheel pictured on page 117). *Complementary colors* are those that lie opposite one another on the wheel—green opposite red, for example, or blue opposite orange, or violet opposite yellow. When used together, complementary colors tend to bring out the best in each other.

Direct complements (red–green) can be used, or *near complements* can be used, that is, one color plus another close to its direct complement, as green with violet-red. *Paired complements* consist of two sets of direct complements, for example, red and green with yellow and violet. *Split complements* are made up of one color and the two colors adjoining its direct complement—orange with green-blue and violet-blue.

Triadic color combinations are made up of three colors equidistant on the color wheel, like yellow–blue–red. This makes a dramatic composition.

Monochromatic composition is the use of several values, tints, shades, and tones of a single color. An *analogous harmony* is the use of only one primary color and its neighbors: blue, green-blue, green and yellow-green; or red, orange, violet, and brown.

If all of the foregoing discussion sounds formidable, remember that such rules are to be studied, assimilated, then bent to the needs of the occasion. Begin slowly and learn to do by doing.

Two-Dimensional Pictures

Flower pictures are sometimes called flower *prints*, but because that term means something else as well (see later in this chapter the section entitled "Decorations"), I shall use the term *picture*. Flowers, leaves, and other plant parts are assembled to form a design featuring the plants for viewing in the same way as a painting, usually in a frame under glass, but sometimes uncovered, as in a wall hanging or plaque.

Designing the Picture

First, decide on four things: the background color, the plants, the shape of the composition, and the frame.

Suppose the plants pressed include buttercups and their foliage, pansies in purple and gold, deep blue larkspur, and medium-blue bachelor buttons. Start with a triangular design to be placed within a rectangular frame. The background color can be white, cream, black, or brown. Arbitrarily choose a creamy-white watercolor paper. Cut a piece to fit the frame and arrange the pressed flowers against that background in an imaginary triangle. Straight lines are few in nature, so the arrangement will be only roughly a triangle.

The buttercup is a tall, slim plant. It can make one side of the triangle. A spray of larkspur can lie along the base. Between them, lay some sprigs of bachelor button. The triangle is becoming a fan. To draw the arrangement together with a focal point, try placing a pansy or two at the point where the stems converge. If two seem to make the arrangement too cluttered, settle for only one in purple, perhaps with a gold center.

Study the picture for balance. Are heaviest masses near the bottom? Yes. Is color balanced? There are blues and purple opposite the yellow, but perhaps that upright stem of buttercups looks a little thin. Try adding one more spray of larkspur just behind it. That does seem to make it more balanced. Blue and purple predominate, sparked with the gold of buttercups. There is some open space to balance the arrangement. The picture seems to invite a companion, a design in the reverse of this one.

Two versions of the larkspur–buttercup–pansy arrangement

The second one could emphasize tints of yellow and orange, accented with blue.

Before coming to a final decision, try arranging those same plants differently. Instead of a triangle, use an oval. Two sprays of larkspur, curved toward one another, will form the two sides, with bachelor button heads scattered among them, and two stems of buttercup laid over them. Place a single small pansy at the top of the oval, a bit off center, and a pair of pansies at the bottom, to add weight. As a final test, lay the picture frame over the whole, and consider whether the elements are balanced. Add or subtract flowers until balance is achieved.

Note that the placement of three pansies forms a slim triangle set against the oval. Check the various criteria of good design, including a balance of color as well as of form. For repetition, blues are repeated in varying tints and shades, the curves of stems echo one another, and most of the flowers are circular in shape. The bilateral symmetry of the larkspur flowers is echoed in the rounder butterfly shape of the pansies. The larger pansy at the base forms the focal point toward which other lines lead. If the pansy at the top of the oval arrangement is distracting, remove it.

Once the picture has been completed, it must be made permanent. If necessary, make a little sketch of the way the plants were arranged, or cut a second piece of background paper and transfer the plant pieces one by one onto it.

Mounting and Framing

Materials for mounting and framing plant pictures can be found at craft supply stores. The materials needed are frames, backing material, glass (preferably nonglare), colorless adhesive (clear glue, white glue, or a latex-based adhesive), small paint brushes, toothpicks, scissors, waxed paper, and, of course, the pressed plants.

To help place the plant pieces correctly, make a very light tracery of dots on the background material. Turn over each fragment of a plant, and dab it lightly here and there with adhesive. A toothpick makes a good applicator, as does a little finger (keeping the other fingers clean for handling the plants). Learn to use forceps to pick up plant pieces, a practice which should lessen the chance of breaking them. Extra glue *can* be blotted up with a damp cloth, but it is even better not to get it there in the first place. It will be enough if the plants are kept from sliding around, since the whole picture will be covered with glass. A petal brushed with glue may tend to curl, so place it with great care. Hold a piece of waxed paper on it as the adhesive sets.

Bit by bit, assemble the plants on the background. Cover the completed picture with waxed paper and place a weight over the whole area until it is dry, or place the picture under glass immediately. The glass should be sparkling clean. Glue a piece of brown wrapping paper or other sturdy paper over the back of the frame to keep out the dust. The act of framing will complete the work. For longest color life, hang it in a cool place away from direct sunlight.

A friendly warning: if work must be left, even briefly, during composing and gluing, cover it against breezes, inquisitive cats, and other hazards with a blotter weighted in place.

Tinting to Maintain Color

There is a way to prolong color in pressed and air-dried plants. Natural color can be reinforced by tinting; for example, white may need to be touched up, lest it gray with age. Apply colors with a light hand. The intent is not to change the flower color, but to maintain it. The flowers should not look painted. Several materials are effective: poster paints, eye shadow, and artist's pastels.

Buy the basic colors in poster paints—red, yellow, blue, white, and black. Yellow mixed with blue will produce green; yellow with red makes orange; and blue with red makes violet. To deepen shades, add a little black. To lighten tints, add white. Add a little of both black and white to produce tones. Poster paints will mix more easily if a drop of detergent is added. Only a small amount is needed at a time. Mix it in a small jar lid.

Apply the color very gently, in small strokes. This will dampen the plants again, so to prevent curling return them to the plant press right away. Lay a piece of waxed paper over them in the press to keep them from sticking to the paper, or to keep the color from coming off.

Artist's pastels come in colors to match nearly every flower. They tend to rub off, but handling the plants gently at every stage will prevent this from becoming a big problem. Spraying the final picture with pastel fixative, obtainable at art supply stores, will prevent rubbing off.

The subtle colors of eye shadow are also effective in maintaining color. Apply them with a soft brush.

Wall Plaques

Not all flower pictures are framed under glass. Others, particularly those using thicker, bolder materials, may be composed as wall plaques.

Begin with a sturdy background material like wood, chipboard, or Masonite. The background is an important part of the design. If the wood itself is to show as the background, prepare it by sanding it smooth and applying shellac according to the directions given on the container, or with shellac followed by varnish. Sandpaper the surface between coats. The wood may be stained before finishing. The wood can be tinted by diluting the artist's oil color with turpentine until it is the consistency of cream, then rubbing it into the wood with a small pad or cloth. When it is dry, apply wax, shellac, or varnish.

The crisscross patchy look of chipboard or flake board is attractive also, when shellacked or varnished. The silvery grain of barn wood makes an excellent foil for blues and reds. Other possible backgrounds are cypress shingles, mat board, canvas board, a shallow wicker tray, place mats, woven mats, or African grass cloth. If the dried materials are chiefly tans and browns, use bright color in the background. You can cover Masonite with another material, perhaps burlap, linen, or a textured cotton. Stretch the fabric tightly over the edges of the panel to the back, and fasten it in place with glue and strong tape. The panel may be framed or not, as best fits the design.

Wooden surfaces can be waxed, although this will cause problems with some adhesives. However, a *hot-glue gun* will take that and other problems in stride. The hot-glue gun is a small, hand-held electrical tool into which a stick of glue is inserted. A pressure on the trigger places a drop of glue on exactly the right spot. It works better if the glue goes on the plant, not the plaque. Place the plant part on the plaque, and the glue sets almost instantly. Burns are possible with any electrically heated tool, so I do not recommend this gun for children's use. Hot-glue guns can be bought at hardware and craft-supply stores.

The elements of color and design will be the same as those recommended at the start of this chapter, Form, color, and texture will be worked with in a somewhat three-dimensional format. Variety of shape, size, and outline will add interest, but as before, too cluttered a design should be avoided. Balance long, spiky elements with round forms.

Choose among the dried materials available. Think of cattails, teasel, milkweed pods, twigs of interesting shape, waxy leaves of magnolia, brilliant bittersweet. Think of pods, cones, seeds, grains like wheat, oats, and rye, and other grasses. Think of the available foliage preserved by the glycerin method, and flowers dried in a dehydrating agent. Driftwood combines happily with other dried plant materials, as do dried fruits, gourds, and fungi.

When beginning work on a plaque, cut a piece of paper or cardboard the same size as the intended plaque. Lay out the design on this, arranging and rearranging until satisfied. Make dots here and there on the plaque to show where to place the plant pieces, then attach them, one piece at a time, using small quantities of strong, transparent glue (Elmer's carpenter's glue is good for most surfaces) or using the hot-glue gun described previously. If the background is a fabric, sew heavier elements of the design to the fabric, using strong linen or cotton thread of a neutral color, threaded through a curved upholsterer's needle. Place the threads where they will be least noticeable, or cover them later with another part of the composition. When the whole plaque is assembled, leave it flat for 24 hours until the adhesive is firmly set.

A few ideas for plaque designs are pictured on page 124.

Cartoons

Pressed plants used to compose pictures representing something else will be referred to here as cartoons. Imagine a parrot constructed from red and yellow tulip petals, sunflower rays for its tail, its clawed feet of fragments of grape tendril, and a beady seed eye. A lady's skirt can be made of the bell portion of a daffodil, with small sunflower-seed shoes peeping out from beneath. Her blouse is made of the daffodil's corona, her rose-petal face will disappear beneath a bluebell hat. The challenge is in trying to find a flower part of just the right shape, resorting to scissors as little as possible.

Cartoons lend themselves to action. Improbable fuzzy lambs made from woolly woundwort (*Stachys lanata*) leap on grass-stem legs over green hills of pressed geranium leaves; or long-legged ostriches of silvery Russian olive leaves, with tails of plumy clematis seeds, chase one another on petiole legs across brownish pressed-leaf plains.

The plants themselves will give you ideas. Their shapes and colors will suggest other figures. Every part of a plant can be used, together or separately—the stamens, the pistils, the leaves, tendrils, or bark. You can make fish, birds, people, and insects such as never were. The cartoons can be mounted under glass or used for greeting cards. This craft is especially good for children.

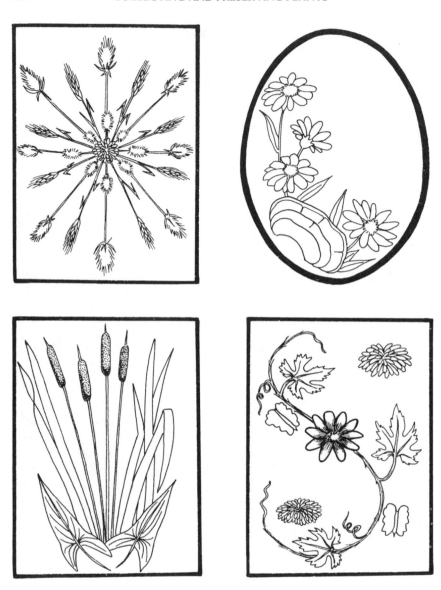

Plaque designs

Seed Pictures and Mosaics

The seed picture or mosaic is a popular art form. Think of the different colors, shapes, and sizes of seeds, and combine them into pleasing designs. Realistic pictures or abstract designs may be created.

Cartoon of a little girl

Just a few of the possibilities are the pale gold of plump wheat grains, the warm brown of spherical mustard seeds, glistening brown flax, tiny black poppy, bright yellow corn-kernel wedges, flat blackish crescents of hollyhock, black-and-white striped sunflower seeds, the soft green of peas and young lima beans, and the white and coppery-red of beans.

First, make some thumbnail sketches, just small drawings in which the elements of the design or picture are worked out, mentally balancing shapes, colors, and textures; consider coloring the sketches. Transfer the design to a suitable background material. If the picture is small, heavy cardboard can form the backing; if larger, use wood or Masonite; or, compose your picture on a plate or platter or other container.

Using white glue having strong adhesive qualities, paint one small area of the background at a time. Sprinkle one kind of seed on this prepared area, or lay the seeds in one at a time.

A variation of the seed picture is one made of spices, and of course these materials can be intermixed.

Two-Dimensional Decorations

The possibilities are endless when using plant parts to form decorative designs. Read through the ideas presented here, and choose one suitable for a beginning project. Combine dried flowers, leaves, and fruits with other materials to make any number of beautiful objects. It is better to practice on smaller things before attempting a larger project. The plant materials will be those that have been pressed flat.

Useful materials to assemble include background papers in varying weights, fragile rice paper, colored construction paper, watercolor sheets, waxed paper, poster board, and mat board. Some frequently used tools are scissors, ruler, pencils, and a paper punch. Other materials include clear glue, clear adhesive tape (Mylar tape is best; other types tend to discolor), clear self-adhesive plastic film or heat-sealing film; and trimmings of all kinds—ribbons, braid, rickrack, passe-partout, colored adhesive tape, decorative cord, and glitter. Most of these materials can be found in art and craft supply shops, dime stores, and hardware stores.

Some possibilities for objects to decorate with plant materials are bookmarks, place mats, coasters, calendars, light-switch surrounds, transparent finger plates, wall hangings, lampshades, fireplace or folding screens, boxes, bookends, trays, stationery, and greeting cards.

Designs, instead of being complete unto themselves, as are pictures and plaques, must fit the shape and function of the object they will decorate. At every stage, work carefully and neatly; measure and make guiding lines with a ruler or template; cut exactly on lines; apply glue with a light hand, and let transparent tape edgings be unobtrusive. The finished product will reflect the care taken at each step, distinguishing an amateur attempt from an artistic result.

Bookmark

Why not begin your designing with pressed plants by making a bookmark? This simple object is a good first project, it can be beautiful, and it is always an appropriate gift for the book lover.

Materials required are transparent plastic film, ribbon about ¼-inch (6 mm) wide, a clear glue, and backing material which can be art paper, blotting paper, or high-quality card stock. Tools will include scissors, ruler, pencil, and paper punch.

Red blotting paper is a good choice, against which plant materials in green, white, black, gold, yellow, and silver would show up well. Cut a piece of the paper 1.5 × 6 inches (3.8 cm × 15 cm). Here is an appropriate place for the S-curve design. Look for a stem in that shape.

A piece of young ivy vine that naturally forms an S-curve, or perhaps the tip of a climbing knotweed would be appropriate. Remember that all the plant parts must be tiny, in scale with the size of the bookmark. The vine will furnish the green color and the major part of the design (red is already in the background). Tiny silver flowers of artemisia, bunched by twos and threes, not quite touching the curving stem, would be attractive.

Once the arrangement has been settled, glue the plant pieces to the backing, using as little glue as possible. Cut two pieces of transparent self-adhesive plastic, each slightly larger than the blotting paper. Peel the protective backing from one piece of the plastic, and lay the paper on it. Peel the second piece of plastic and lay it over the top, making sure to lay it down without wrinkles the first time. There is no second chance. Trim the plastic so it projects about ¹/₁₆ inch (2 mm) on all sides of the background paper. Complete the bookmark by neatly punching two holes at the top and drawing through them a bit of narrow ribbon.

An alternative to the kind of transparent plastic described is the heat-press method given earlier (see Chapter 6).

Place Mats

The method for making place mats closely follows that for making bookmarks, except that everything is on a larger scale. You will need background material, clear glue, and plastic film. Because of the difficulty of handling a large piece of self-adhesive plastic film that must be peeled from its backing, the heat-press kind of film is better here.

For each place mat, cut a rectangle of backing material 12 × 18 inches (30 cm × 45 cm). As you compose your design, remember the future placement of dishes on the mat. A design that will not be covered by the dishes is best. Try a C-curve in the lower left corner, balanced by a smaller design in the upper right. A possible combination would be a creamy-white water-color paper for the background, and an arrangement

A bookmark

of scarlet cardinal flowers and blue larkspur against a curving larkspur stem and its leaves. In the opposite corner place a small grouping, one flower of each kind and a leaf.

When the flowers have been glued in place on the backing material, cut a piece of plastic film large enough to fold about the paper mat, extending a little on each side. The plastic bonds best to itself, so the overlap is needed. Place it in the heat press for the length of time recommended by the makers of the press. Weight the place mat while it cools to prevent curling. Trim the plastic leaving $1/16$ inch (2 mm) overlap on all edges, rounding the corners.

Place mats make beautiful, useful, durable, easily cleaned additions to table decoration. Store the mats away from the light when not in use, to preserve their color.

The mat described uses flowers of small to medium size. Bolder designs can be used here, perhaps a pressed single pink rose against a triangle of fern. Arrangements of a shape other than a C-curve would be appropriate. A set of six place mats could be designed, each featuring the same flower combination, or six different flowers; or a set decorated with spring or fall leaves from six different kinds of trees; or of woodland

A place mat and coaster

flowers of spring; or of six different ferns; or arrangements featuring different seasons, or different habitats (bog, woodland, meadow, stream, rocky area, and lakeshore).

Coasters

To match the place mats, surely you should have some coasters, which can move comfortably to the living room for the dual function of decoration and protection of furniture surfaces.

Enclosed within its circle, the design can be a single large flower, a set of three flowers with a leaf or two in the background, or a little nosegay. Try not to let the flower pedicels project awkwardly to one side. Let the design radiate from the center, a single cluster of Queen Anne's lace backed by a bit of its own ferny leaf, against a background of blue, perhaps, or an ivy stem twined into a circle against white. Add a single primrose in the center. Combine three bluebells, against pale yellow, and cover their pedicels with one of their own leaves, or a round leaf from a ground mallow. Most ferns would be too large for a coaster, but parts of fronds can be snipped off and used.

Laminate the design and its backing in plastic film to preserve the coaster for long use. This would make a good project for children, using perhaps six different autumn leaves. Make sure the plant materials are not so thick as to make the final result uneven, or a glass placed on the coaster may tip.

Calendar

Another use of the plastic-laminated flower design is a calendar. The flower composition can be like that of a flower picture.

Materials needed are a background material, clear glue, transparent plastic film, ribbon or cord, a calendar, small or large, and a stiff backing material large enough to carry both the flower picture and the calendar.

Assemble the flower picture and glue it to the background. Imagine a collage of ferns against a deep cocoa-brown background. The ferns could be placed as though all were growing from the base of the picture, up from shining golden-brown oak leaves; or perhaps laid crisscross to fill the entire background in a more abstract design.

Laminate the completed design within plastic film and trim the edges. The picture can be framed with colored adhesive tape. Attach the picture to the stiff backing of poster board or mat board with ribbon or decorative cord, extending the ribbon for hanging the whole. Center and carefully glue or tie the calendar in the space beneath.

Light-switch Surrounds

Introduce a lovely personal touch into home decoration by making flower-decorated light-switch surrounds.

Purchase transparent plastic plates to fit the switches. Cut a piece of background material colored to match the decor and build the design within this rectangle, balanced about the center.

If the colors are blue and pale yellow with touches of burgundy, any one of those three colors could form the background, but let us imagine the pale yellow there. Select from among your stock of pressed flowers. Larkspur gives a rich blue; for burgundy, use snapdragons. Both kinds of flowers should be small ones. Bits of curved stems plus tiny leaves of privet will provide additional materials.

Use four shallow S-curves, one each at the top, the bottom and the two sides. Along these curves, in graceful balance, arrange the other elements of the design. Use materials sparingly, not to make a clutter. Another variation on the same color combination would be blue and yellow Johnny-jump-ups, against a burgundy background. Dainty materials in other colors are alyssum, buttercups, tiny roses, lady's bedstraw, and quaking grass.

When you have glued the plants to the background, carefully fasten the paper inside the transparent light-switch surround, taping all the way around the edge with transparent Mylar tape. Fasten the whole assembly to the switch box with the screws provided. Every time the switch is used, you should experience a feeling of pleasure at the sight of an ordinary object transfigured to one of beauty—and uniquely your own.

Finger Plates

You can protect and decorate an area on a door with a finger plate.

At a hardware or department store, purchase a transparent plastic or glass plate made for the purpose. Cut a background sheet just slightly smaller than the plate. Against this sheet build the design of flowers. An elongated oval might be an effective design shape, or a softened rectangular outline that could be filled in an all-over pattern, almost like a tapestry. Buttercups interspersed with sprays of forget-me-nots, against a dusty rose background make a possible combination, or green-and-white variegated ivy and pale gold foxtail grass heads against black for a more formal design. Think of *your* favorite colors and seek out plants to carry out your theme.

Once the design is set, glue it to the background. Attach the assembly to the back of the finger plate with transparent Mylar tape and fasten the plate to the door with the screws provided.

Wall Hanging

A variation of the finger plate will produce a wall hanging. Frame the plate with passe-partout or colored adhesive tape. Instead of attaching the completed plate to a door, thread ribbon or decorative cord through the top mounting holes for hanging it up. Thread a matching ribbon or cord, with a tassel or a medallion, at the bottom to complete the design.

Lampshades

Imagine a lampshade decorated with your favorite flowers glowing away of an evening, bringing summer into a winter room.

You will need a plain lampshade, clear glue, and clear self-adhesive plastic film.

The design can encircle the shade, be placed on one side, or be repeated, perhaps three times. Its colors will be influenced by artificial light. Yellow and orange are least affected by artificial light. White also changes little. Pink or red takes on a yellow cast. Blue and violet may be dulled. It is a good idea to look at the flowers to be used under indoor lights, so that the finished product can be envisioned. The design will be more important than the color, since it will often be seen in silhouette. Keep it simple and uncluttered, and use leaves and stems of definite outline.

Either glue the design to translucent art paper and cover it with plastic film, then attach the whole unit to the lampshade; or glue the flowers directly to the shade, then cover with transparent film. The first method is safer since it gives you a second chance without ruining a whole shade. Encircle the shade with braid at the top and bottom as trim. If the shade is tall and narrow, a vertical design will be best. A short, broad shade would look best with a circular design, or one that goes all the way around. Three possible combinations of plants are white-poplar leaves combined with sprays of white alyssum; sweet-gum leaves and sprigs of wild fall asters; or red Japanese maple leaves with coral bells.

Fire Screen or Folding Screen

A screen provides a chance to use bolder designs composed of pressed plants. Looking at the outline of the whole screen, decide whether to build in a series of squares like individual tiles, or give the illusion of a single whole plant growing up through a panel, or intermingle plants in a collage with some pieces lying crosswise, or make an overall pattern. Consider whether to assemble the screen in such a way that the flower-decorated parts could be replaced if they fade. The fire screen is *not* intended as a heat shield, but to be placed before the fireplace when it is not in use.

A folding screen will be made of three wooden frames, one, two, or three of which will be filled with transparent plastic or glass. The design can be embedded between the pieces of the plastic or glass, then fastened into the frame, or the designs can be placed on a background of clear or translucent plastic film, and affixed to the existing glass with clear glue.

Imagine central stems of any vine, the pieces assembled to look like a single upright stem. Along this, place individual pressed flowers and leaves in a simple design to achieve the effect of a single flowering branch.

The delicate fronds of pressed seaweed would look especially lovely mounted in a flowing vertical design in a screen. Some seaweeds are too delicate to glue directly. When collecting, look ahead to this decorative use and float some directly onto a transparent material, glass or plastic.

Folding screen, lampshade, and light-switch surround

Large blossoms like poppy, clematis, African daisy, rudbeckia, or daffodil could be spread up the rectangle of the screen in a repeating pattern, interspersed with fern fronds; or ferns or grasses alone could form the pattern.

This ambitious project should reward you with a striking individualistic expression of your love of plants.

Boxes, Bookends, Trays, and Other Containers

Many objects can be converted from the ordinary to the artistic by the application of pressed-plant designs. Possible subjects for this art form are a wooden jewelry or treasure box, a pair of bookends, a tray, desk accessories, or wastebaskets. Other items will come to mind.

Wooden boxes can be purchased unfinished from craft supply stores, or may be constructed. Sandpaper all surfaces until they are satin smooth. Vacuum or wipe away any particles. For best results, use a tack cloth. Apply a coat of clear lacquer (available at paint supply stores or craft shops) according to the instructions given on the container. When that is dry, sand or steel-wool the box lightly again, remove the dust, and apply a second coat. Be sure to treat the back of the wood also, to prevent warping.

Arrange the flowers on the *wet* surface of the second coat and press them down gently with a brush. When they are thoroughly dry, apply another coat. Two coats may be necessary to cover thicker or more uneven flowers.

Real flowers will have a liveliness and detail not found in painted ones. This technique may be used to beautify a pencil holder, a cigarette box, bookends, and a wastebasket as a matched set to use with a desk at home or in an office. Choose someone's best-loved flowers or favorite color combination. Wall plaques of flowers lacquered to wooden panels are another variation.

Children can learn to make gifts, applying this method, using containers made of wood, cardboard, or even metal. In the fall, the children could gather flowers, leaves, and grasses, press them, store them in old telephone books; then at Christmas they would have materials at hand for making gifts.

Instead of lacquering the flowers directly to the object, try building the design first on a paper background, then glue on the whole assembly. Then apply coats of lacquer, or cover the design with a piece of self-adhesive plastic film, trimmed neatly with rounded corners.

Rejuvenate old wooden or metal trays by sanding and refinishing them with a first coat of paint or lacquer, depending upon the surface, and applying flower designs.

Stationery and Greeting Cards

Watch for and press the very tiniest plant parts for this purpose, for example, both the blue flowers and the interesting fruits of forget-me-not. Many small weeds, like shepherd's purse, stellaria, silvery cinquefoil, medic, sweet clover, and alyssum have delicate flowers, leaves, and fruits. Grasses are especially lovely, as are fragments of ferns. Press individual flowers and leaves or very small sprigs for this use. Many trees, such as maple and elm, have small flowers. Separated from their twigs, these flowers press well.

Designs should be restrained and simple, little fans or circles or S-curves of flowers, leaves, or grass panicles.

Buy boxes of plain stationery to decorate, or buy attractive paper in bulk at a craft supply store and cut it to the desired size. Standard typing paper of good bond quality, folded into quarters, makes an attractive size. The base paper should be of good quality, to be worthy of the work decorating it. Stationery sizes acceptable to the U.S. Postal Service are from $3\frac{1}{2} \times 5$ inches (8.5×12 cm) to $6\frac{1}{3} \times 11\frac{1}{2}$ inches (15×28 cm) for a 1-ounce weight.

NOTE: In countries having different specifications, locally standard sizes would be used.

Glue the plant parts to the first page of folded stationery. Be sure each part is fastened down, but avoid excess glue. The glass-plate method of gluing (see Chapter 5) works well here.

Package the stationery attractively and use it as gifts to favored friends, as sale items at a bazaar, or to use for your own correspondence. Stationery makes a fine gift or sales item, not only because it is useful and attractive, but because it gets used up, and more can be given out or sold.

To package stationery, make a folder of lightweight cardboard or art paper, designed to fit it exactly. First make a model of plain paper, following the sketch on page 136 to be sure everything is right. Use this as a pattern and transfer the design to heavier paper. Before folding, score the cardboard by marking the fold line with a blunt table knife. It will then fold neatly, right on the line. If it is necessary to construct a box, consult the directions in Chapter 6. Folders and boxes can be decorated too.

The plant-decorated stationery can be left plain, or the fragile flowers can be protected by covering the outside with a fold of extremely thin rice paper. The design will show through the rice paper, taking on a misty quality. To hold the assembly together, punch two holes in only the front half of the stationery and thread through a short length of narrow ribbon. The flowers could be covered with a piece of rice paper, cut slightly larger than the design, and glued on. Mix 1 part of white glue with 3 parts of water and gently work the glue into the rice paper with a flat brush. The glue penetrates the paper and fastens it to the stationery. Allow the sheets to dry. If they wrinkle a bit, when they are dry press them under weights until they straighten out. Another way to protect the plant decorations is with a piece of self-adhesive clear plastic, sold in department stores and craft shops, but this will give a less delicate effect.

Card shops often have leftover envelopes they will sell, or a package of envelopes can be purchased and the cards designed to fit them. Envelopes can be made without difficulty. Study how one is made and adapt its design as necessary. Foldover type stationery, complete in itself, with the flower design glued *inside*, is another possibility.

Facial Tissue/Waxed-paper Technique

A special technique using facial tissues, waxed paper, and pressed flowers, makes elegant stationery. The materials needed are scissors, scalloping or pinking shears, narrow satin or felt ribbon, waxed paper, white glue, facial tissue, stationery or notepaper, envelopes to match, pressed flowers or seed heads, a watercolor brush, and a paper punch.

Cut a piece of waxed paper the same size as a facial tissue. Dilute the glue 1-to-1 with water. Using a watercolor brush, paint the glue onto the

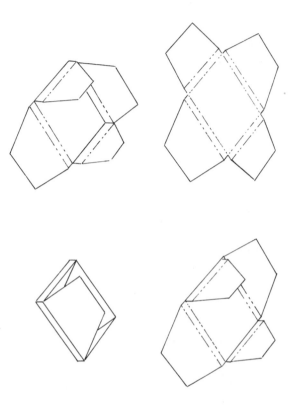

Making a folder for stationery (drawn by Richard MacFarlane)

waxed paper, back and forth until it is covered. The glue may tend to ball up. Place little flowers or seed pods on the glue so they will decorate only the *front* of the finished card. If desired, sprinkle on some glitter with a salt shaker or from pinched fingers.

Lay *one ply* of the tissue over the flowers. Paint on more glue, lightly. If too much is used, the tissue will disappear, but you must get it covered. Lay the waxed paper and tissue aside to dry, and when it is dry, turn the paper face down between seamless layers of brown paper (a grocery sack will do) and press it with an iron set at "cotton."

Fold the waxed paper in half, tissue side out. The flower design should be facing up. Place a piece of notepaper, white or tinted, inside the folded waxed paper. Punch holes either through all four thicknesses or just the first two. Slip a ribbon through the holes but don't tie it. Cut around three sides of the stationery with the scalloping or pinking shears, then scallop or pink the flap of the envelope also.

Use a prepared sheet of stationery to cover the lid of a box, place a diagonal ribbon on it, and use the box to package five or six pieces of stationery and envelopes to match. Or make folders or boxes as suggested previously for other stationery.

Greeting Cards

The thought of greeting cards for special occasions and holidays follows closely upon that of stationery. With a little planning ahead and some enjoyable time spent in design and assembly, you can produce your own beautiful and individual greeting cards. Plant cartoons (see earlier in this chapter), as well as a wide variety of designs, lend themselves to this use.

Stationery decorated with pressed plants

For Valentine's Day, hearts, flowers, doilies, and ribbons will combine to send your message. Look up the language of flowers and choose appropriate blossoms to say what you mean. Kate Greenaway's *Language of Flowers*, first published in 1884 and recently republished by Avenel Books will help you. Sweet alyssum means "worth beyond beauty," and the flowers are a nice size for stationery. Of course, you must tiptoe among the meanings, not to send the wrong message. For example, blue violets mean "faithfulness," while yellow violets mean "rural happiness," and lavendar, "mistrust."

Materials to assemble will include construction or art paper in red, pink, and white, transparent glue, lace, ribbons, doilies, and tiny flowers and leaves.

Cut symmetrical heart shapes by folding a piece of paper lengthwise and cutting half a heart. Make patterns in several sizes and trace around them. Either build your designs on rectangular stationery or have heart-shaped cards. This is a time for sentimental design. Just one of the infinite number of possibilities is a pink paper heart mounted on a white linen-finish card. On the heart, fasten a doily and in the middle of the doily glue a nosegay of pressed forget-me-nots and bleeding hearts, backed by a fragment of fern; or scatter little florets directly on a heart and back that with a doily. Think of dainty coral bells and single aster flowers. You can make folded cards and tie them with lace or ribbon; or cartoon lavender-and-lace ladies by using pink or lavender tulip petals, Queen Anne's lace, and real lavender blossoms, which will scent the card. Many plants, basswood, lilac, morning glory, and philodendron, have heart-shaped leaves.

Christmas Cards

For Christmas cards, be as restrained or as lavish as you wish. Sprigs of real balsam fir, culled from your Christmas tree, will give off the scent of the holiday, as will a piece of arbor vitae (white cedar), commonly grown as an ornamental shrub. Tiny twigs of flat-needled yew or hemlock will be useful in designing these cards, but spruce needles are too stiff and thick and would drop off, so avoid them.

Make a tree of the tip of a hemlock branch, trimmed with tiny coral bells, artemisia, and purple alyssum, with a white stellaria star at the top. Draw a swath of glue across the card with a brush and sprinkle the wet surface with glitter. Perhaps, in a reversal of the usual form, write your greetings on the front the card, and glue the flower design inside to protect the fragile pieces.

Other Holidays and Special Occasions

On other holidays, think of shamrock leaves for St. Patrick's Day; flowers of St. John's wort for Midsummer's Day; red, white, and blue flowers (cardinal flowers, white everlastings, and bachelor buttons) for July 4th; pressed Chinese lantern for Halloween; grasses and fall leaves for Thanksgiving. For any special day, send a card bearing the favorite flowers of the recipient. Lilies of the valley could decorate a sympathy card.

Flower and Leaf Prints

A technique that can be used for making stationery or greeting cards or for decorating other objects is that of making flower and leaf prints. Many leaves and flowers can be used as stamps for printing. Only perhaps eight or nine can be done from each leaf or flower because it is fragile. To prepare, collect materials in shapes that will print well and press them in the usual way.

Use poster paint (with a drop of detergent to make it spread better) or colored ink. Make test prints at first to find out how much paint or ink to use. Paint an even coat of ink or paint on one side of the leaf or flower with a soft paintbrush. Turn over the leaf and place the inked side on the card or paper to be decorated. Cover the leaf with a piece of blotting paper or paper towel and smooth over the blotting paper by hand or using absorbent tissue, with quick, even strokes. Remove the blotter and lift off the leaf carefully, not to smudge.

A single perfect leaf will make a graceful print, or make overlays of several leaves or flowers in turn, perhaps using different colors. At every stage, take great care not to make smudges. A variation of this method would be to print your own gift-wrapping paper using flowers or leaves in a repeating overall pattern on plain tissue paper or newsprint.

Flower and Leaf Prints on Fabrics

Using sheets of dye paper, iron the dye first onto the leaf or flower, then place the plant on the fabric to be decorated and iron it again, working always between sheets of protective paper. Not all fabrics will take prints well. The fabric must have more than 35 percent polyester content. The technique will not work on 100 percent cotton, wool, silk, or rayon. Some manufacturers market special dyes for making prints on fabrics. One such dye is marketed in kits called "Nature Prints," using "DRY-DYE," sold by:

Unicolor, Inc.
Champaign, Ill. 61820

or sold at craft supply shops.

The methods listed in the previous pages are only some of the possible ornamental uses of plants. I urge you to select a project, practice with scrap materials to learn the technique, then enjoy making objects that express your own ideas and creativity.

10

Ornamental Uses
—Three-Dimensional

Air-dried plants or those dried in a dehydrating agent, or in glycerin, will have retained their three-dimensional shapes. Here at hand is a wealth of material for creating bouquets, centerpieces, wall plaques of greater depth than those made from pressed flowers, arrangements under glass, corsages, and other decorations. See Appendix 2.

The principles of design in three dimensions are similar to those given for flat compositions, but you must be able to view them from all sides. Dried flowers follow the rules for the arrangement of fresh flowers.

Bouquets

Design

Underlying patterns among which to choose are the crescent, the S-curve, the cone, the fan, the globe, the ovoid, the pyramid, the L, and vertical and horizontal arrangements. As a general rule, the basic lines of the composition should be approximately one-and-a-half times the height or width of the container. Dried arrangements may look better if they are a little taller than that.

Balance color and bulk so that darker colors and a heavier massing of materials are near the base. There should be a focal point where the eye comes to rest, provided by the largest or brightest flower, by the largest massing of small or light-colored flowers, or by some unusual shape. Other lines of the design lead to it, or appear to radiate from it. Consider scale, using materials neither too large nor too small for the container.

You cannot do better, when learning to arrange flowers, than to study the ancient Japanese art of flower arranging. Some of their work is

S-curve and vertical arrangements

very formal, and to learn it you will need to study the subject intensively. On a less rigid basis, we can borrow some of their underlying principles. Traditionally, Japanese flower art represents some phase of natural life. Therefore, all kinds of plants—grasses, leaves, flowers, branches, and bark—may be harmoniously combined.

The triangle underlies all Japanese flower compositions. The tallest line of the composition is called the "heaven" line, the one of intermediate height is called "man," and the shortest, "earth." Seen from above, these three lines spread away at angles to each other. Secondary lines, "mountain" and "meadow," are added. Both face toward the heaven line and stand at two-thirds and one-third of its height, respectively. Additional plant pieces, or "helpers," may be used to fill in, but place the main lines first and always keep them in mind. Those elements of the design reinforcing the heaven line should face toward it; those reinforcing the earth line may droop over the edge of the container.

In very formal arrangements, no part of the flowers, stems, or foliage may touch the container. Stems are secured by a wedged stick holder; a piece of twig is split, the stems are inserted into it, and the twig is wedged into the container opening.

The use of the crescent in flower arrangement

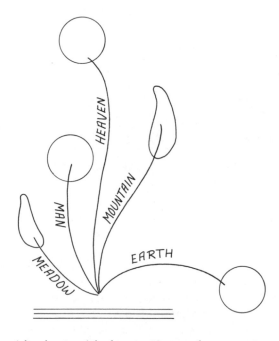

A line drawing of the elements of Japanese flower arranging

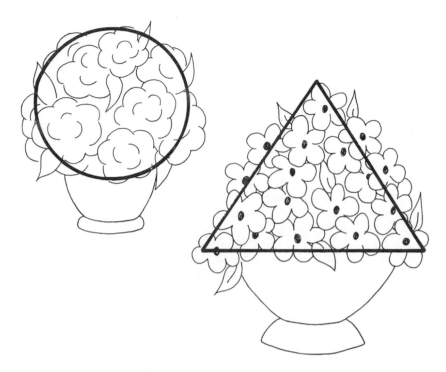

Arrangements based on the globe and the pyramid

It is possible to find the three basic lines on a single stem, but more often three separate flowers or stems are used.

While the Japanese style gives a more formal, balanced but asymmetrical arrangement, the plants available may call for the more informal ovoid or globe-shaped groupings that were favored in colonial America. Although the flowers are massed, there still must be a balance of light against dark, with the dark closer to the base and the center of the bouquet, the light colors at the top and periphery. The stems of small or light-colored flowers may be bundled together, secured with a rubber band or wire to a pointed stick, and then added as a group to the composition.

Securing and Supporting Items

Once a general plan has been established for the bouquet, the mechanics of building a stable arrangement must be considered. No matter how delicate it appears, underneath there must be enough weight and bracing to keep the bouquet from tipping over and its parts from slipping or falling out.

Some of the materials that can be used to promote stability are sand, glass beads, Plasticine clay (not children's modeling clay), styrofoam, melted paraffin, needlepoint holders, florist's sticks, florist's green and brown tape, pipe cleaners, wire, wire hairpins, and Mylar tape. Stems may be inserted into the holes of glass "frogs," or through the meshes of hardware cloth or crumpled chicken wire. Scissors, a tape measure, watercolor brushes, and a pointed stick for punching holes will be required.

Fill an upright container with sand, but work out some way to seal in the sand when the bouquet is complete, because an accidental tipping will spill it. In a low arrangement, use Plasticine, styrofoam, melted paraffin, needlepoint holders, or a frog to anchor the stems. Sticks and wire will lengthen or reinforce stems, while pipe cleaners will make bendable ones. Cover stem reinforcements with florist's tape. In the finished bouquet, cover all artificial aids in some way, perhaps with bits of plants or leaves, or with rocks.

Crumpled 2-inch mesh chicken wire set over 2 or 3 inches of sand in a container will provide many little openings into which to insert the plant stems. Styrofoam is one of the easiest materials to use. It can be cut with a knife to fit any need, and holes poked in it to hold the stems.

Containers should be chosen to form an integral part of the arrangement. Oval or round vase shapes complement circular designs; tall containers emphasize vertical lines; low ones are best with horizontal lines. A container that rests on the floor to hold large branches or plants should be heavy in both fact and feeling, to be in scale and to counterweight the plants. Watch for unusual containers at garage sales; those made of china, glass, silver, ironstone, wood, or metal are all suitable. Pitchers, sugar bowls, goblets, bowls, vases, and old lamps may be used. Among suitable kitchen utensils are butter bowls, enamelware, and utensils of aluminum, pewter, brass, copper, and iron. Consider trays, baskets, shells, and coral. Watertight containers are not needed for dried plants. Try to select those holders of simple outline, little ornamented, in order to direct attention toward the flowers.

Possible Plant Combinations

By judicious choosing and collecting of plants, the whole range of colors, and all their shades, tints, and tones will be available. Work formally in careful color combinations, or combine materials wildly; among plants, almost any color grouping works.

Imagine a few combinations: rosy-red prince's feather, white daisies, coral bells, and a branch of red maple leaves dried in air; or white poplar leaves and golden tansy sparked with purple wild marjoram heads; or dried white peonies, blue salvia, blue larkspur, white baby's breath, and a few stems of green-and-white orchard grass.

Suppose that you have at hand the following collection of dried plants.

Air-Dried

Tansy (gold)
Wild marjoram (purple)
Pearly everlasting (white)
Strawflowers (rose, burgundy, yellow)
Maple leaves (pale yellow)
Queen Anne's lace, "bird's nests" (pale green)
Quaking grass (tan)
Chinese lantern (orange)
Baby's breath (white)

Glycerined

Peony leaves (silvery bronze)
Barberry leaves (red)
Russian olive leaves (yellow)
Maple leaves (olive green)

Dried in Dehydrating Agent

Gladiolus (rose, burgundy)
Bells of Ireland (pale green)
Larkspur (blue)
Maidenhair fern (pale green)

What combination should you use for an arrangement to enliven the gray-white days of post-Christmas?

One grouping could be essentially purple and gold. Make your first (heaven) line with a branch of pale yellow sugar maple leaves, air dried on their twigs. Pour 4 to 6 inches (10 to 15 cm) of sand into an earthenware bowl of medium height. Thrust the base of the maple branch into the sand. Make two lines at angles to this with golden tansy, for man and earth. Insert two clusters of purple wild marjoram at the mountain and meadow heights. Seen from above these elements will spread away from each other to make a well-rounded bouquet. Fill in the spaces with glycerined Russian olive leaves in yellow.

Another grouping could be an ovoid arrangement in a pale green vase, using light green, rose, burgundy, and white. Here, try using the elements in threes. First, set the height with three Queen Anne's lace bird's nests in pale green, then three stalks of Bells of Ireland. So far the arrangement looks stiff and drab, with only those straight elements in pale green, but now place six rose-colored strawflowers at heights about half that of the first six elements. If necessary, extend the length of their stems by taping the stems to wire.

The arrangement still looks thin, dominated by the vase, so try redesigning it by adding three gladiolus stalks, two with rose-pink flowers, one with burgundy. Make these the same height as the bird's nests. The shape of the gladiolus will echo that of the Bells of Ireland, the colors will repeat those of the six everlastings. Add three burgundy strawflowers near the base of the arrangement. It is now complete,

Dried plant arrangement

balanced in color and shape. For a more lavish arrangement, fill in the outline with dainty white baby's breath.

When every element is exactly where you want it, stability can be insured by pouring ½ inch to 1 inch (1.5 to 2.5 cm) of melted paraffin over the top of the sand.

Make a fall arrangement in orange, tan, and green, using orange Chinese lantern for the dominant elements, combined with tan quaking grass. The flattened grass heads will repeat the lantern shapes. Fill in with pale-green maidenhair fern. Moving to the blue side of the color spectrum, try blue larkspur dried in a dehydrating agent, combined with silvery-bronze glycerined peony leaves and masses of white pearly everlasting dried in air, for a cool, delicate effect. For a Christmas arrangement, departing from the usual, compose a bouquet of red glycerined barberry leaves, pale-green Bells of Ireland, and white baby's breath.

These are only a few suggestions of the possible ways to design an arrangement. When you have before you your own collection of preserved plants, your imagination will help you to produce an arrangement reflecting your own tastes.

If you make up several bouquets at one time, while you have your materials spread out, you will have them on hand when needed for gifts, for decorations, and for sale.

Bouquets should be placed away from direct sunlight and away from open windows or a humidifier. They can be kept fresh by touching them up with a lightly dampened brush or cloth, or with a glycerin-and-water solution in the case of glycerined foliage. Remove any pieces that have become damaged and replace them with new material.

Earlier in the book (Chapter 7), instructions were given for tinting dried plants to retain their natural color. Some people paint seed pods and flower heads for more dramatic color. Others think that all materials should be natural. My own opinion is that you should do whatever works for you. You are the artist. If poppy capsules sprayed with purple paint are just what you need for your composition, do it.

Use a spray booth when spraying any plant materials with paint, varnish, or clear plastic, because paint drifts. Unless confined, it will soon coat everything in the room. A large cardboard box set on edge makes an effective spray booth. Either work through the flaps of the top, left on the box for the purpose, or make a curtain for the front. Materials like goldenrod heads in seed can be stabilized or colored by spraying them with clear varnish or plastic, or with paint. Goldenrod tops, painted green, make miniature trees for model layouts.

To dip plants in paint, float some oil paint—gilt, silver, or colored—on top of water in a jar and dip the plants through it. This requires less paint and makes a thinner product, less liable to injure the plants.

You can also apply colors with a brush. Types of paints to use are quick-drying enamel; water-soluble casein or latex interior paint; poster paint (the easiest to use since it dries in an hour or less); and artist's tube paint, thinned. While still wet, plants may be sprinkled with glitter, if desired.

Centerpieces

A centerpiece should be low, built with horizontal lines in a shallow container, so that people seated around a table can see over it. A low round bowl is suitable. In it build a horizontal arrangement of three or six larkspur stalks, three rose zinnias in the center, and massed xeranthemum flowers in rose, lavender, and white, all inserted into a styrofoam base. A different arrangement could be a low styrofoam mound in a silver tray, completely hidden by zinnia flowers in yellow and orange. An old-fashioned Thanksgiving centerpiece could be a wicker cornucopia set in a woven reed tray, spilling out corn, gourds, nuts, seed pods, dried lemons, oranges, and pomegranates.

A centerpiece

Wall Plaques

Wall plaques were described under "Two-Dimensional Design" in Chapter 7. Think of plaques in three dimensions; the design factors are the same, the plaque will simply project more from the background.

Arrangements Under Glass

The arrangements just described are larger ones that stand exposed to the open air of the room. Think also of building plant combinations protected inside of glass containers. The bridal bouquet dried in borax and sand and displayed under a glass dome is a traditional example.

Find glass containers of pleasing shapes—globes, domes, bell jars, apothecary jars, candy jars, and even those in which grocery items are packaged. Select plants in scale with the size of the container, and appropriate to its shape. The smaller types of everlasting flowers will be better here, such as globe amaranth, xeranthemum, and pearly everlasting. Use delicate grasses, bits of dried fern, yarrow, baby's breath, and the leaves of privet and boxwood.

A half-sphere of styrofoam makes a good stabilizer. Plan the arrangement, then poke holes in the styrofoam and insert the flower and

Dried roses inside a glass globe

leaf stems. A hot-glue gun will cement them firmly in place, if necessary. Use enough filler plants and foliage to hide the styrofoam.

To develop ideas, first sketch the container, then draw a series of little thumbnail sketches until a workable idea develops, or work directly with the plants. Keep a used piece of plastic foam, poked full of holes, for experimenting.

Just two suggestions would be: one full-blown rose and two rosebuds, dried in borax and sand, combined with glycerined rose leaves in a glass globe; or white daisies, purple globe amaranth, baby's breath, and boxwood leaves in a low candy jar.

Corsages

Corsages can be made of dried plants quite as well as with fresh flowers. A corsage should be about 6 or 7 inches long (15 or 18 cm), with few components.

Imagine a corsage of three pink rosebuds, dried in dehydrating agent, with glycerined peony leaves and rose-pink xeranthemum.

To make a corsage, begin with one element. Wrap the stem (or the wire that has been inserted in place of a stem) with green florist's tape.

Add a second flower and wrap its stem to the first one with tape. Add each element until the corsage is assembled. Make sure that each stage is completed neatly. Finish by tying on it a bow of narrow ribbon.

Corsages can be designed for special holidays. For Thanksgiving, try nuts and acorns, fastened with wires slipped through holes drilled with a fine drill, combined with stalks of wheat or a corn tassel. For Christmas, use little cones of spruce or hemlock, supported on slender wires wrapped through their lower scales. Combine them with sprigs of fresh white cedar or balsam, or with a feathery amaranth head, and tie on a bow of red or green velvet. The cones can be gilded or brushed with thin glue and sprinkled with glitter.

Other Three-Dimensional Decorations

Many other beautiful and unusual decorations can be made besides the bouquets and other uses of dried plants described above. The traditional bowls of fruits, gourds, and nuts, and the Indian corn tied together with its own husks and fastened to a door as a greeting to visitors, are familiar to us all. There are many, many more ideas.

Christmas Decorations

The midwinter holidays call forth our most elaborate decorative designs. Wreaths of evergreen boughs and ropes of white cedar have long been familiar. Here are ideas for some decorations that can be made of dried plant materials.

Cone Pendant

To decorate an entrance door, make a cone pendant. Cut 1-inch-wide ribbon (2.5 cm) into varying lengths. Use red and green velvet ribbon if you like the traditional colors, or any combination that suits you. Fasten a pine cone, plain or gilded, to the lower end of each length of ribbon, using fine wire or glue from a hot-glue gun, and tie the upper ends of the ribbon together securely with wire, raffia, or strong cord, leaving a loop for hanging. Make a lavish ribbon bow to cover the tied upper ends and hang up the pendant for a holiday greeting.

Garlands

To make a lovely garland to drape over an arched doorway, a mantel, or up a staircase, use a double thickness of clothesline twisted loosely together. Tie it at intervals with raffia or strong cord for stability.

Use nuts and pinecones, silvered or gilded or plain, plus seedpods, the rosy-red tufts of celosia and pieces of white cedar or balsam (most other evergreens would soon lose their needles). Fasten the design elements to the rope with wire, or glue them on with a hot-glue gun, in a repeating pattern. You can intersperse artificial materials like sleigh bells or ribbon bows if you like, or you can keep the whole garland in natural materials. Tufts of goldenrod heads, dipped in bright red or green paint, or the gilded button-like heads of tansy would work in handsomely here.

Christmas Tree Ornaments and Miniatures

Shining glass balls and tinsel do not make the only possible decorations for Christmas trees. Everyone knows about strings of popcorn and cranberries, but from among the dried plants you have assembled come other possibilities. Teasel heads dipped in green, red, gold, or silver paint and sprinkled with glitter will hang like miniature trees upon the larger one. Hang gilded half-walnut shells, fitted with gold wire handles and filled with tiny strawflowers. Cover styrofoam balls with dried blossoms, some of them fragrant. Small nuts and seed pods can be glued to slender ribbon swags to loop among the branches. Use bouquets of multicolored strawflowers, a cluster of red sumac berries, cones hung singly or in clusters, made brilliant with gilt paint and glitter.

Is there a dollhouse in the house? It needs a Christmas tree. Use a teasel head, painted green and sprinkled with glitter or cake decorettes. Make ornamental plants for the dollhouse by painting goldenrod or aster heads green and standing them in clay in little pots. Make a vase filled with tiny strawflowers or pearly everlastings. Make landscape plants for the outside of the dollhouse by using larger goldenrod or tansy tops, or those of other appropriately scaled plants, painted green. As suggested earlier, those same trees and shrubs would add to a model train layout.

Wreaths

The wreath has long been a favored holiday decoration. To make a wreath, buy preformed wire or styrofoam bases or make the bases of slender branches of forsythia, privet, lilac, willow, or grapevine. Clip away the side twigs and bend the branches into circles of the size desired, perhaps 15 inches (38 cm) in diameter. If the branches are stiff, soften them by soaking them in cool water for several hours or overnight. Overlap the ends 6 inches (15 cm) and tie them securely. Lay several of these circles, one upon another, staggering the joinings, and tie them together firmly at several places with wire or strong string. Weight the wreaths as they dry to keep them from warping.

When the frame has dried, add coarse sphagnum moss (obtainable from garden-supply houses) for filling, and cover the frame with florist's paper or waxed paper, or wind it with strips of cheesecloth or other cloth of a neutral color. Both moss and paper can be held to the frame by wrapping it with a spiral of lightweight string. Insert the stems of dried plants into this base or glue them on. If all the parts are to be assembled by gluing, you might use the branch from the frame without wrapping.

Build your wreath upon the frame, beginning with heavier elements, perhaps pinecones, in a repeating pattern. Among these, glue (or wire or tie) cones of smaller size and different shapes, some on their sides, some set on end. Nuts may be interspersed with the cones. You may wish to cut some cones into crosswise slices for their flowerlike effect. Such a wreath of cones and nuts has a rich, fruitful look. You may gild some of the elements, spray the whole assembly with gold paint, or leave them in their own rich natural browns. Complete the wreath with a lavish bow.

Dried plant wreaths can be made in the following combinations or others of your own devising:

- A mingling of glycerined leaves and dried fall leaves for an autumn wreath.

- Russian olive, barberry, and beech leaves interwoven with white cedar for subtle color.

- Artemisia and pearly everlasting combined with bayberries and a silver bow for a frosty look.

- Colored statice, pastel-painted cones, and strawflowers for a pretty effect.

- Clusters of red sumac berries alternating with olive-green glycerined privet leaves and feathery red celosia, in a muted version of the traditional red and green for Christmas.

- Dock dried at several color stages—green, pink, and russet—combined with lavender heather or wild marjoram.

- Gray-white artemisia for a basic color, sparked with rose celosia and strawflowers in white, pink, and burgundy.

- A wreath of herbs combining sprigs of fragrant dried lavender, sage, wild marjoram, basil, rosemary, and thyme, in gentle colors.

A child's version of the wreath can be assembled on a cardboard back. Glue on small spruce and hemlock cones with white glue, then paint them with gilt paint. Fasten a loop of gold string to the back for hanging, either in the child's room or on a Christmas tree.

Wood Arrangements

Wood itself can be used to create striking, unusual arrangements, alone or in combination with other plant parts.

Keep alert for unusual wood forms, perhaps a contorted twig, or a gracefully curved one, or driftwood polished and bleached by sun, water, and sand. Peel the bark from a dead branch and examine the wood beneath, which may be smooth and shining, or interestingly decorated by the tunnelings of bark beetles. Look for the texture of the bark itself, on dead wood. (Do not peel bark from a living tree or that tree will die.) Some bark is smooth, ringed horizontally, or speckled with lenticels; some is checkered, ridged, or coarsely grooved or chunky.

All wood used for ornamental purposes is referred to as *driftwood* by some artists, although *decorative* would be a better word. Some kinds like red manzanita from the West Coast, or twigs of corkscrew willow, can be purchased in florist or craft shops.

Wood that you gather yourself will almost always need to be cleaned. Begin by scrubbing it with a brush and soap or detergent and water, until all dirt and loose particles are removed. Cut away any

A wood arrangement

decayed parts with a knife. If a very smooth surface is desired, rub the wood with sandpaper or steel wool. When it is thoroughly clean and dry, apply a coat of shellac, mixed according to the directions on the can, as a sealer. To retain a natural look, sand lightly after the shellac dries. Wax, varnish, tint, or stain; anything from shoe polish to household bleach may also be used. To add a little color, rub in a small amount of well-thinned paint with a cloth or fingers.

Make use of the natural shape and texture of the wood, either as a background for other plant arrangements, or displayed for its own design.

Compatible containers for wood are metal, pottery, and other wood. Dried fungi, as well as the colored, odd-shaped pebbles that we gather compulsively on a beach or any rocky place, will combine well with wood.

Make a permanent background for the display of fresh or dried plant arrangements by fastening a chunky or branched piece of wood to a tray, and tucking little containers here and there, disguising them with sphagnum moss or glycerined foliage.

As just one suggestion for an arrangement, imagine three zigzag redbud branches representing heaven, mountain, and man, a C-curve of grapevine (meadow), and an earth line of grapevine, set in the cleft of a bare, polished chunk of applewood.

Let the wood itself suggest to you possible arrangements and uses.

Fragrant Plants

Flowers and leaves that retain their fragrance in drying have special uses as sachets, potpourris, pomander balls, and as herbs for cooking and for making wreaths. See Chapters 2 and 6 for other instructions.

Gather the petals of roses, lavender, honeysuckle, orange and mock orange, spice pinks, and other fragrant flowers, and the leaves of mint, scented geraniums, and the herbs basil, rosemary, thyme, marjoram, tarragon, lemon balm, and lemon verbena. The leaves of tansy have a strong odor at first, but as they dry they mellow and have a pleasing fragrance. Tansy is reputed to repel moths. The scented geraniums are delightful to grow, coming in such scents as apple (*Pelargonium odoratissimum*), peppermint (*P. tomentosum*), lemon (*P. crispum* or *limoneum*), rose (*P. graveolens*), and nutmeg (*P. fragrans*). All these geraniums will grow as house plants.

Sachets

Make small bags or pillows of dainty but porous cloth, like thin cotton, about 2 by 3 inches (5 by 8 cm). Fill them with dried materials and either sew them shut or tie each with a bit of ribbon or lace. Often their fragrance will last for years, where they lie tucked among clothing or linens. Crushing the packet lightly, occasionally, will renew the scent.

Pomander Balls

The old-fashioned pomander ball is an apple or an orange stuck full of whole cloves, rolled in spices, and enclosed in a net covering decorated with a ribbon bow and a sprig of some fragrant herb like rue. The ball may be hung in a closet or bedroom, or placed among clothing in a drawer. The fragrance lasts for years.

A newer version, sometimes called a herb ball, uses a styrofoam sphere for a base. Fasten to it a collection of dried herb leaf clusters, either inserting them into holes poked in the foam, or using a hot-glue gun to glue them on. Intersperse small colorful dried flowers and petals among the herbs. Attach a loop of ribbon and hang the herb ball above a dressing table or desk, where every passing air current will release its fragrance.

Potpourri

To make a potpourri, select an attractive container with a tightly fitting lid. Place a layer of dried petals in the jar. Sprinkle the petals lightly with salt and a combination of equal parts of orrisroot, spices, and brown sugar. Orrisroot, obtainable at a drugstore or a natural foods store, serves as a fixative for the petals, preventing spoilage. The spices are a blend of allspice, cinnamon, cloves, mace, or nutmeg. Place another layer of petals and leaves in the jar, and more of the condiment mixture. Repeat until the jar is nearly filled. Stir and mix thoroughly. Age the blend 3 weeks, stirring it twice a week. Once blended, the fragrance will last for years. Every time you remove the lid from the jar, the fragrance will escape to bring a whiff of summer to your room.

Some people recommend adding a few drops of rose oil or geranium oil and of glycerin (all should be available at a drugstore). Continue adding dried petals and leaves from time to time, if you so desire. It is a good idea to keep a careful record of what materials you did include, to aid in repeating an especially pleasant result, or in avoiding the repetition of a mistake.

For every suggestion offered in the foregoing chapter, there are many more possible uses of dried plants. I urge you to experiment, invent, read other books on specific techniques, and sign up for courses taught by craft shops.

Dried Plants

Bauzen, Peter, and Suzanne Bauzen. 1982. *Flower Pressing*. Little Craft Book Series. New York: Sterling.

Booke, Ruth Voorhees. 1962. *Pressed Flower Pictures*. New York: Avenel Books. Out of print, but look for it in libraries.

Eaton, Marge. 1973. *Flower Pressing*. Minneapolis, Minn.: Lerner Publications.

Scott, Margaret K., and Mary Beazley. 1980. *Making Pressed Flower Pictures*. Batsford, U.K.: David and Charles.

Squires, Mabel. 1958. *The Art of Drying Plants and Flowers*. New York: Bonanza Books. Excellent but out of print.

Underwood, Raye Miller. 1952. *The Complete Book of Dried Arrangements*. New York: M. Barrows. Out of print, but worth looking for.

Appendix 1

Table of Abbreviations

bu, bushel, bushels
°C, degrees Celsius (Centigrade)
cm, centimeter, centimeters
DBH, Diameter at breast height
°F, degrees Fahrenheit
fl oz, fluid ounces
ft or ft., foot, feet
g, gram, grams
gal, gallon, gallons
in, in., inch, inches
kg, kilogram, kilograms
km, kilometer, kilometers
l, liter, liters
m, meter, meters
mi, mile, miles
ml, milliliter, milliliters
mm, millimeter, millimeters
oz, ounce, ounces
p., page
pp., pages

PDB, paradichlorobenzene,
  a fumigant
pk, peck, pecks
pl., plural
pt, pint, pints
qt, quart, quarts
R., range
S., section
sing., singular
sp., species (singular)
spp., species (plural)
sum, summer
T., township
T, tablespoon, tablespoons
TRS, township, range, and section
tsp, teaspoon, teaspoons
var., variety
vars., varieties
wntr, winter
yd, yard, yards

Table of Equivalents

Linear

10 mm	1 cm	12 in	1 ft
100 cm	1 m	3 ft	1 yd
1000 m	1 km	1728 yd	1 mi
1 mm	0.039 in	1 in	25.4 mm
1 cm	0.39 in	1 in	2.54 cm
1 m	3.28 ft	1 ft	0.30 m
1 m	1.09 yd	1 yd	0.91 m
1 km	0.62 mi	1 mi	1.61 km

Fluid

1 ml		0.03 oz	1 oz		29.57 ml
1 liter		1.06 qt	1 qt		0.95 liter
1 liter		0.26 gal	1 gal		3.78 liter
1 tsp	1/6 oz	4.9 ml	12 T	¾ C	177.4 ml
3 tsp	1 T		16 T	1 C	237 ml
1 T	½ oz	14.8 ml	1 C	8 fl oz	237 ml
2 T	1 oz	29.57 ml	2 C	1 pt	474 ml
4 T	¼ C	59 ml	4 C	1 qt	945 ml
8 T	½ C	118 ml	4 qt	1 gal	3.78 liter

Weight

1 g	0.03 oz	1 oz	28.35 g
1 kg	2.20 lb	1 lb	0.45 kg

Volume

8 qt	1 pk	4 pk	1 bu

Temperature Conversion

To convert temperature readings in degrees Fahrenheit (°F) to degrees Celsius (formerly Centigrade) (°C), or vice versa, use one of the following two formulas.

Degrees Fahrenheit to Celsius:
(°F −32°) × 5/9 = °C
Example:
(80° F−32°) × 5/9 =
48° × 5/9 = 240/9 = 26.6°C

Degrees Celsius to Fahrenheit:
(°C × 9/5) + 32° =
(225/5) + 32° =
45° + 32° ± 77°F

Appendix 2

Plants to Gather or Grow for Ornamental Purposes

The plants listed here are those most useful and reliable to gather or grow for ornamental purposes. Almost any plant, however, will yield parts that can be used in pictures, designs, or arrangements.

The abbreviations used in the following table are:

ann, annual
bien, biennial
blk, black
brn, brown
cult, cultivated
fl, fls, flower(s)
fr, frs, fruit(s)
fragr, used for fragrance, as in
   potpourris
glyc, glycerin
grn, green
hang, hang to dry
imm, immature
lf, leaf
lvs, leaves
mat, mature
peak, peak of flowering

per, perennial
pict, picture
plnt, plant
prpl, purple
sp., species (singular)
spp., species (plural)
st, sts, stem(s)
3-D, three-dimensional
   arrangement
2-D, a flat picture or other flat
   design
wht, white
yel, yellow
lav, lavender

Plants to Gather or Grow for Ornamental Purposes

Common Name	Scientific Name	Type	Part Used	Color	Time to Gather	Treat-ment	Use	Remarks
Acroclinium	*Helipterum* spp.	cult ann	fl	white pink rose yellow	imm	hang	3-D	
Baby's breath	*Gypsophila paniculata*	cult ann or per	fls	white rose	fls partly open	press hang agent	2-D 3-D 3-D	One of the most useful
Bittersweet	*Celastrus scandens*	per vine	fr	orange-red	after frost	hang	3-D	Protected, but you can grow your own or buy at farmers' markets or florists
Celosia

See also Cockscomb, Prince's-feather | *Celosia plumosa* | cult ann | fls | red yellow orange rose purple | peak | hang | 3-D | Long lasting, vivid color |

Common Name	Scientific Name	Type	Part Used	Color	Time to Gather	Treatment	Use	Remarks
Chinese lantern	*Physalis alkekengi (franchetti)*	cult per	fr	orange-red	peak	hang	3-D	A favorite in arrangements; once established, long-growing
Cockscomb	*Celosia cristata*	cult ann	fl heads	red, pink, rose	before peak	hang	3-D	Long lasting, vivid color
Columbine	*Aquilegia* spp.	wild or cult per	fls	blue, yellow, white, pink, red, purple	peak	press	2-D	Lovely spurred fls
Coral bells	*Heuchera sanguinea*	cult per	lvs	green	peak	press	2-D	
			fls	pink	peak	press	2-D	
				red		agent	3-D	
				white		air-dry	3-D	
						press	2-D	
			lvs	green	peak	glyc	3-D	
Cucumber, wild	*Echinocystis lobata*	wild or cult ann	fls	white	peak	press	2-D	
			lvs	green	peak	press	2-D	

Cupid's dart	*Catananche caerulea*	cult per	fls	silver & blue	imm	hang	3-D	One of the best everlastings
Delphinium	*Delphinium* spp.	cult per	fls	blue lavender white purple	peak	press agent	2-D 3-D	One of the easiest to dry; holds color

Ferns

Sensitive fern	*Onoclea sensibilis*	wild or cult per	frond fruiting stalk	green dark brown	peak fall winter	press dries in place	2-D 3-D	Almost all ferns can be used ornamentally when pressed or dried in air, unless they are too large
						hang	3-D	
Cinnammon fern	*Osmunda cinnamomea*	wild per	fruiting stalk	cinna-mon brown	summer, when first formed	air-dry	3-D	Spray with fixative
Ostrich fern	*Matteucia struthiopteris*	wild or cult per	fruiting stalk	dark brown	summer	air-dry	3-D	

APPENDIX 2

Common Name	Scientific Name	Type	Part Used	Color	Time to Gather	Treatment	Use	Remarks
Globe amaranth	Gomphrena globosa	cult ann	fls	white pink rose purple lavender	imm	hang	3-D	
Forget-me-not	Myosotis spp.	wild or cult ann or per	fls lvs frs	blue green brown	peak peak peak to mat	press press press	2-D 2-D 2-D	Good for dainty designs
Goldenrod	Solidago spp.	wild or cult per	fls frs	gold white	imm past peak	press air-dry air-dry	2-D 3-D 3-D	Spray with fixative to retain white fluffy look, or with green paint for miniature trees
			frs	tan	after seeds fall	air-dry	3-D	Dry calyxes attractive

Common name	Scientific name	Type	Part	Color	mat/peak	air-dry	3-D	Notes
Gourds	*Cucurbita* spp.	cult ann vine	frs	orange yellow green white	mat		3-D	
Grass								
Bluegrass	*Poa pratensis*	wild or cult per	lf, stem & head	pale green to tan	peak	press hang	2-D 3-D	
Honesty	*Lunaria annua*	cult ann or bien	frs	silver	mat	hang	3-D	Rub off covering membranes
Hydrangea	*Hydrangea* spp.	cult shrub	fls	lavender white blue pink green	peak	agent hang	3-D 3-D	
Joseph's coat	*Amaranthus tricolor*	cult ann	fl heads	red	peak	hang	3-D	
Larkspur	*Delphinium* spp.	cult ann	fls fl spikes	blue lavender red white purple	peak	press agent	2-D 3-D	A top choice; holds color well and petals may be used separately

Common Name	Scientific Name	Type	Part Used	Color	Time to Gather	Treatment	Use	Remarks
Lilac	*Syringa* spp.	cult shrub	fls	lavender white pink purple	before peak	press agent air-dry	2-D 3-D fragr	Also use separate florets
Love-in-a-mist	*Nigella damascena*	cult ann	fls	blue pink	peak	press	2-D	
			lvs	green	peak	press	2-D	
			capsule & involucre	green	peak	hang	3-D	Inflated capsule and finely cut involucre unusual in bouquets
Maple	*Acer* spp.	tree						Maples are versatile; the flowers, leaves, and fruits of all maples may be treated by the methods given for Norway maple
Norway maple	*A. platanoides* (var. *schwedleri*)	tree	fls	yellow-green	peak	press	2-D	
			lvs	green maroon	peak or imm	press	2-D	
						air-dry glyc agent	3-D 3-D 3-D	

							Treat in the same ways as leaves taken earlier
				maroon	in fall		
Red maple	*A. rubrum*	tree	frs	green maroon	peak	See above	3-D
			fls	reddish	imm	See above	
			lvs	green	peak	See above	
				scarlet	in fall	See above	
Sugar maple	*A. saccharum*	tree	fls	pale green	imm	See above	
			lvs	green yellow	peak in fall	See above	
Marjoram, wild or cult	*Origanum vulgare*	wild or cult per	fl spikes	purple	peak or past peak	press	2-D
						hang	3-D
					peak	air-dry	fragr
			lvs	green	peak	air-dry	fragr herb
				yellow	in fall	press	2-D

Common Name	Scientific Name	Type	Part Used	Color	Time to Gather	Treatment	Use	Remarks
Pansy	*Viola tricolor*	cult bien	fls	yellow blue purple brown maroon	peak	press agent	2-D 3-D	One of the best flowers to use
			lvs	green	peak	press	2-D	
Pearly everlasting	*Anaphalis margaritacea*	wild per	fls	white	peak	press	2-D	One of the best for small white flowers
Pinks	*Dianthus* spp.	cult ann or per	fls	white red pink maroon	peak	press agent air-dry	2-D 3-D fragr	
			lvs	gray-green	peak	press agent	2-D 3-D	
Poppy	*Papaver* spp.	cult ann or per	fls	white red pink orange	peak	press	2-D	Silky petals useful in many ways

Common name	Scientific name	Type	Part	Color	Stage	Method	Dimension	Notes
			lvs	green	peak	press	2-D	
			capsule	green brown	mat or imm	hang	3-D	
			seeds	black	mat	air-dry	seed pict	
Prince's-feather	*Amaranthus hybridus* var. *hypochondriacus*	cult ann	fl head	red	peak	hang	3-D	Spectacular plumes
			lvs	green red	peak	press	2-D	
Privet	*Ligustrum* spp.	cult shrub	fls	white	peak	press	2-D	All parts useful
						agent	3-D	
						air-dry	fragr	
			buds	green	imm	agent	3-D	
			lvs	green	peak	press	2-D	
						agent	3-D	
			fr	black	mat	hang	3-D	
Queen Anne's lace	*Daucus carota*	cult or wild per	fls	white	imm	hang	3-D	Delicate
					peak	press	2-D	
			lvs	green	imm to peak	press	2-D	Ferny appearance

Common Name	Scientific Name	Type	Part Used	Color	Time to Gather	Treat-ment	Use	Remarks
			fr	green to tan	bird's-nest stage	hang	3-D	
Rose	*Rosa* spp.	wild or cult shrub	fls	white red yellow rose	½ open	press agent	2-D 3-D	Single types Among loveliest when dried
			petals		peak	air-dry	fragr	
			lvs	green	peak	press agent glyc	2-D 3-D 3-D	
Starflower	*Scabiosa stellata*	cult ann	fls	tan	mat	hang	3-D	Star-like centers
Statice (Sea lavender)	*Limonium* spp.	cult per	fls	yellow lavender blue white rose	peak	hang	3-D	Foolproof material for drying

Stocks	*Mattiola* spp.	cult ann or per	fls	white red, pink yellow purple	peak	press agent air-dry	2-D 3-D fragr	
			lvs	gray	peak	press agent	2-D 3-D	
			frs	gray	imm	hang	3-D	
Strawflower	*Helichrysum bracteatum*	cult ann	fls	white red yellow pink bronze	imm—1/3 to 1/2 open	hang	3-D	Bright colors that last
Sumac	*Rhus* spp.	wild shrub	frs	green	early imm	hang	3-D	Colorful, dramatic
				rose	imm	hang	3-D	
				red	peak	hang	3-D	
				maroon	mat	hang	3-D	
			lvs	green	peak	press	2-D	Dramatic
				red	in fall	press	2-D	
			fl stalks	green to red	peak	air-dry	3-D	Strip of fruit, striking lines
Xeranthemum	*Xeranthemum annuum*	cult ann	fls	white red pink	imm	hang	3-D	

Index

NOTE: Page numbers in italics indicate illustrations.